The Anthropocene Project

The Anthropocene Project

Virtue in the Age of Climate Change

Byron Williston

OXFORD
UNIVERSITY PRESS

OXFORD
UNIVERSITY PRESS

Great Clarendon Street, Oxford, OX2 6DP,
United Kingdom

Oxford University Press is a department of the University of Oxford.
It furthers the University's objective of excellence in research, scholarship,
and education by publishing worldwide. Oxford is a registered trade mark of
Oxford University Press in the UK and in certain other countries

First Edition published in 2015
Impression: 1

Published in the United States of America by Oxford University Press
198 Madison Avenue, New York, NY 10016, United States of America

British Library Cataloguing in Publication Data
Data available

Library of Congress Control Number: 2015933285

ISBN 978–0–19–874671–3

Printed and bound by
CPI Group (UK) Ltd, Croydon, CR0 4YY

For George and Pippa

Preface

Imagine a group of socially polished and prosperous dinner guests who, after dining handsomely, find themselves unable to leave the room they are in. The windows and doors are unlocked and there are no guards anywhere preventing departure. They all want to leave but somehow cannot do so. And so they stay for days, pacing the single room, arguing over trivialities, grooming themselves mindlessly, until their fatigue gets the better of them and they turn violently on one another. This is the strange plot of Luis Buñuel's film, *The Exterminating Angels*. It is, among many other things, a biting commentary on the manner in which social and economic privilege can constrict the imagination. These people are certainly frustrated by the inertia gripping them and yet—this is the key—they don't consciously pine for anything very specific outside these walls. The constricted space holding them is the outward expression of their own cramped imaginations. They are, to this extent, quite unlike the literal prisoner who thinks longingly of his girl, his kids, sitting on the beach with a cold beer.

Buñuel's film should make us deeply uncomfortable as we contemplate the link between our prosperity and the crisis of climate change. All the evidence tells us that in refusing to place meaningful constraints on our greenhouse gas emissions we are allowing for the emergence of a terrible future. Of course there's a sense in which many of us want things to be going otherwise, but in our posturing about this issue we are unfortunately pacing a room built and maintained by the fossil fuel energy regime, with our consent. One of the things obscured from view in this space is the toll our inaction is going to exact on the vital interests of future people. I say obscured from view rather than expunged altogether because our exploitation of fossil fuels has caused a weird split in us. We know that our overuse of the stuff is wrecking the future climate, but the sheer kinetic power it gives us is mesmerizing. The spell can be broken only if we develop what Heilbroner calls a 'collective bond of identity' with people of the future. The shift to a decarbonized economy—our primary ethical task in the short term—won't happen without this more fundamental alteration of our temporal consciousness.

Those who benefit most from perpetuation of the fossil fuel regime characteristically present it to us as a necessity. This is unvarnished ideology, which means that the problem we face is inescapably political, but it is crucial to see that it is also moral-psychological. That is, we should recognize, with Adorno, that we exist in a 'social totality' defined by capitalism's productive forces. These forces shape our desires and beliefs as much as our major institutions. I don't think we're as far gone as Adorno apparently does—because we're not devoid of the requisite convictions and values—but his analysis is illuminating. Debunking ecologically destructive and alienating power structures on the one hand and restructuring the relevant subset of our mental states on the other are two aspects of the same moral-political task. This is why it is simplistic and self-serving to attribute all the blame for climate change to, say, oil company executives. There's more than a grain of truth in their perennial apology that in seeking to squeeze the earth dry of every available drop of fossil fuel they are only catering to our demands. This is a hard truth, which is why we have become so clever at not fully noticing the real shape and scope of the mess we are in. Just like Buñuel's dinner guests we must grasp the fact that our metaphorical doors and windows are not locked. I hope this book helps us to accomplish this.

By comparison with most books on climate ethics by philosophers this one is meant to be accessible to a broad readership. The problem of climate change is currently being worked on from a bewildering variety of perspectives, both within the walls of the academy and beyond them. I think that a good deal of the confusion experienced by many people about this phenomenon is owing to the profusion of ways of talking about it. So I have tried to synthesize a sizable chunk of this information. Though harnessed to a single argumentative purpose I draw liberally on the relevant insights of journalists, climate scientists, sociologists, psychoanalysts, politicians, historians, novelists, activists, security specialists, genocide scholars, and more. I'm a trained philosopher, and in the kinds of questions it raises as well as the approach taken to answering them the book is squarely philosophical. But in my view the very last thing we academics should be doing when discussing climate change is indulging our tendency to write exclusively for fellow specialists. Wherever possible I have therefore avoided technical verbiage. At many points technical issues do arise, but to the extent that these become parochially philosophical I have consigned them to the notes. I hope this enhances the flow of the text and contributes to the larger aim of drawing concerned and intelligent people together so that we might come up with a coherent and effective way of thinking about our expanding crisis.

I owe thanks to many people for helping me to develop the ideas in this book over the years. Martin Schönfeld originally got me thinking about climate change nearly fifteen years ago, while I was still a devoted and happily insular scholar of early modern philosophy. I'm not sure I would ever have come around to this issue, and to environmental ethics more generally, had it not been for those innumerable late-night discussions with him. He planted the seed that has become this study. I have also been encouraged in my approach to this problem by two of the towering figures in climate ethics, Dale Jamieson and Henry Shue. Not only have both of them read and responded helpfully to my work but, more importantly, in their own writings on climate change they have defined and opened up an entire field of study. Despite the pleasure and excitement of staking out some ground on *terra incognita* it is very difficult to do philosophy in an area in which virtually no literature yet exists, but that is what Jamieson and Shue (among a handful of others, including Simon Caney and Stephen Gardiner) have done.

Simon Dalby read the portions of the manuscript dealing with the securitization of climate change, and I thank him for his expert advice in this area. My colleagues and students at Wilfrid Laurier University, the University of Waterloo, and the Balsillie School of International Affairs have been listening to me talk about climate change for a few years now. I have benefited from their insights into some of the stickier problems the issue raises, both philosophical and policy-related. I have also benefited immensely from the patient and thorough comments provided to me by the anonymous readers for Oxford University Press, and I thank them warmly. Dominic Byatt at the Press deserves a special note of thanks. This book's scope is ambitious, but Dominic supported it intrepidly and enthusiastically right from the start and guided me smoothly through the early and harrowing stages of the publication process. Finally, I thank my partner and our children. Without their love and support I'm not sure I could have sustained so many years of work on such a bleak problem. That I close on a note of hope and by pondering the possibility of intergenerational forgiveness is due in no small measure to their salubrious influence on my thinking and general outlook.

Contents

1

Introduction

Climate Change and the Virtues

1.1 Introduction

The 2014 Fifth Assessment Report (AR5) of the Intergovernmental Panel on Climate Change (IPCC) includes a detailed assessment of climate change's likely impacts on human security. Summarizing these findings the chairman of the IPCC, Rajendra Pachauri, stated:

> If the world doesn't do anything about mitigating emissions of greenhouse gases and the extent of climate change continues to increase, then the very social stability of human systems could be at stake. (Quoted in Semeniuk 2014)

This book is an extended reflection on the ethical meaning of this threat to our civilization. I begin this chapter by exploring an illuminating historical analogue to our current crisis: the global experience of climate change in the seventeenth century. The goal here is to draw a contrast between the general reaction to climate change at that time and the very different reaction many had to the Lisbon earthquake roughly a century later, when the principles of the Enlightenment had become established in European culture. The juxtaposition describes a salutary historical trajectory. Given our own inaction on climate change we need to ask if we are advancing the enlightened understanding of such events or if we will repeat the more tragic set of responses that dominated in the seventeenth century. Next, I will motivate the virtue-ethical approach to this problem, beginning with an introduction to the three virtues whose analysis forms the core of this book—justice, truthfulness, and hope—and going on to say why it is that these three virtues are so important in the age of the Anthropocene. Finally, since climate change is bound to alter profoundly the material conditions in which our social lives are embedded—that is, our situational realities—I present a brief critical engagement with the so-called situationist challenge to virtue ethics. I conclude that situationism does not represent a mortal threat to the virtues (nor to virtue ethics), as some of its proponents suggest it does, though it does offer some genuine insights into the moral life, insights that are especially useful for the climate crisis.

1.2 Stage-Plays, Sodomy, and Sorcery

In the chapters to follow I'm not going to offer an historical analysis of climate change's impacts on human collectives, but rather lay out as clearly as possible the climate path we are currently on and the threat this path represents to the web of human interrelations we have established in the age of globalization. So I will be talking about what our best science can tell us about the unfolding future, though my principal target is how we should understand all of this as presenting a specifically ethical task. But, despite the firm focus I will take on the present and future it can be helpful to at least begin by examining an historical analogue for the period of uncertainty and breakdown into which we seem to be stumbling. The seventeenth century—specifically 1618–88—experienced dramatic climate change as well as terrible wars, rebellions, and revolutions. Recently, Geoffrey Parker has put the two phenomena together in a persuasive historical synthesis. A brief look at his account of the period's struggles will pay valuable dividends as we try to gain a fuller picture of our own challenges.

The period in question is a segment—in fact the climatic nadir—of the Little Ice Age. During this time there was a decrease in solar energy, coupled with an increase in volcanic and El Niño activity. This led to anomalously cool global temperatures, significant enough to cause widespread disruptions to established weather patterns and the human systems that are geared to them. As just indicated, the central climatic event here was the El Niño, now known as the El Niño Southern Oscillation (ENSO). ENSO events, which affect mainly the equatorial Pacific Ocean, have knock-on effects throughout the global climate. For example, when temperatures cool, there is a shift in the normal distribution of air pressure in the Pacific. Whereas normally it falls in the West and rises in the East, the reverse is now the case and this leads to a corresponding shift in winds, from Asia to America. This in turn causes massive rains to fall in the Americas but can diminish dramatically the monsoon rains that are so vital to Asian agriculture. Probably because of a slowdown in the ocean conveyor, Europe can see colder and drier conditions, especially in the north. And it *was* dramatically colder there in the Little Ice Age. As Goldstone notes, 'at one point even the Bosporus froze solid, creating a land bridge from Europe to the Middle East that had not existed for millennia' (2013, 36).

Societies around the globe are structured to climatic normality from which ENSO events are a departure. Most societies are capable of adapting to the roughly five-year cycle of these events. But as Parker explains, the seventeenth century experienced them twice as often, placing enormous strain on adaptation measures. The reason for this frequency was a vicious climatic feedback involving three elements. First, because of a diminishment of sunspots, the amount of solar radiation hitting the earth was reduced, lowering average global temperatures. This phenomenon is referred to as the Maunder Minimum, and it can set up ENSO events. Second, the latter look to have caused an increase in volcanic eruptions. The way this works is that the ENSO causes a massive displacement of water (some 24 inches) from the Asian to the American coast. This in turn puts increased pressure on tectonic plates just where the world's biggest and most active volcanoes can be found. The result is more eruptions, spewing sulphur dioxide into the atmosphere. Third, since sulphur dioxide particles reflect sunlight (when they bond with water vapour to form sulfuric acid) temperatures are lowered even further, continuing the cycle (Parker 2013, 16–17).

Globally, there were scores of revolutions, rebellions, and wars between 1618 and 1688. In Europe alone, Spain, France, Sweden, Poland, the Dutch Republic, the Holy Roman Empire, and the British Isles were at war for virtually the entire century. Take the British Isles as an example. Although they were relatively peaceful until about 1620 (unlike almost every other part of the continent), the rest of the century was not so quiet. In this period Britain was at war with Spain (1625–30), with France (1627–29), there was civil war in Ireland (1641–43) and in England (1642–48), a war with Scotland (1650–51), with the Dutch (1652–54), the Spanish (1655–59), and twice again with the Dutch, in (1665–67) and (1672–74) (Parker 2013, 27). The same pattern shows up beyond Britain. In China, warfare became commonplace in the Ming–Qing transition (1619–83) (Parker 2013, 29). Much the same can be said for Russia, the Polish–Lithuanian Commonwealth and most of the Ottoman Empire and its vassal states, stretching across north-eastern Africa, the Middle East, Crimea, and southern Eurasia.

The most serious problem across the globe was the decline in agricultural productivity consequent on severe, persistent drought conditions, though increased flooding was also an issue (especially in the Americas). The social and personal effects everywhere were pronounced. Everyone felt the pain, but the hardest hit were those who were already socially marginalized. Slavery, for instance, increased dramatically. In China, Europe, and Africa, millions of people were enslaved, while in Russia many more were reduced to serfdom. The religiously heterodox everywhere were persecuted and suicide amongst women who had been raped or left destitute by war increased. The details are truly gruesome, as is the scale of the atrocities (Parker 2013, 668–9).

I want to draw three main lessons from the seventeenth-century episode of climate change, mainly about the social and political responses to it. First, as I have just indicated, Parker's work is important because of the causal links he makes. Climate change can have severe impacts on human systems, especially agricultural systems. These impacts, in turn, can affect sociopolitical systems, at the extreme by inducing war and revolution. And finally, the latter can have profound impacts on (a) what we now call human security; and (b) back on the natural environment itself, as groups consume material resources with no regard for the ecological impact of such consumption. However, putting things this way vastly oversimplifies the nature of the causal links involved and suggests, falsely, that such links are deterministic. In Chapter 3, I will therefore expend considerable effort in showing how complicated all of this can get.

Talk of causal chains between climate and human security brings me to the second point. In the seventeenth century there were different ways of responding to the climate crisis. In broad strokes, governmental authorities responded either through intelligent adaptation aimed at securing the welfare of subjects or through policies which stupidly or recklessly inflamed tensions among people. Japan in the Kan'ei era (1624–43) was a good example of the first strategy. Like almost everywhere else this period was marked in Japan by famine and consequent peasant revolts. But Tokugawa Iemitsu (the third Shogun of the Tokugawa dynasty) responded by building more granaries, increasing the sophistication of the communications infrastructure, creating appropriate economic legislation, and refusing to engage in aggrandizing warfare (Parker 2013, 695).

Unfortunately, the more common response was to turn opportunistically to war and plunder:

> [I]n East Asia, the repeated harvest failures caused by adverse weather in the early 1640s had two dramatic political effects. First the famines and popular rebellions in Jiangnan fatally weakened the Ming as they struggled against the inroads of "roving bandits" from the northwest. Second, drought and cold in Manchuria so reduced harvest yields that the Qing leaders concluded that invading China offered the only way to avoid starvation. (Parker 2013, 676)

Often, similar governmental responses to the crisis, though less dramatic on their faces, could have far-reaching and disruptive social consequences. For example, in the teeth of severe famine in the British Isles, Charles I significantly exacerbated tension by insisting on saddling Scotland with a new liturgy in 1637. Similarly, in Catalonia in 1640 royal troops desecrated the churches of their opponents, helping to fuel an eighteen-year-long rebellion (Parker 2013, 677). Again, this response—governments acting in ways that inflamed tensions among groups—was typical around the world. Indeed, at least according to Parker, Japan was virtually alone in responding to the crisis in a manner that did not increase human misery.[1]

The third and most important lesson has to do with one prominent way people tried to understand what was happening. It has to do with what Parker calls 'peccatogenetic' explanation, that is, explanation of events that points to their putative origin in human sin. As Parker explains, 'stage plays, sodomy and sorcery'—the salient set of behaviours said to be consequent on sinful or vicious inner states—were in fact less frequently invoked to explain catastrophe than stars, eclipses, earthquakes, comets, and sunspots, but they were nevertheless often pointed to, with predictably baleful effects. You can't torture a sunspot but you can get away with hanging someone suspected of witchcraft, especially after a crop failure or other natural disaster. For instance, in Germany cold temperatures following a hailstorm in 1626 brought about the arrest, torture, and murder of 900 people accused of sorcery (Parker 2013, 9–10).

In sum, and again with the exception of the Japanese, people in the seventeenth century were catastrophically wrong about both what was causing the climatic chaos and what should be done about it. Universally, they misunderstood the causes of the changes assailing them while generally responding to those changes in ways that exacerbated their negative effects. These events took place on the cusp of the European Enlightenment, so it is worth comparing them to what happened after the great Lisbon Earthquake in 1755. The earthquake triggered a crisis in how we understand large-scale geophysical events. Many thinkers of the time refused to drag God and sin into their explanation of the earthquake. After all, as Martin Schönfeld points out, the day of the earthquake was a Christian holiday (All Saint's Day) and many of the dead had been in church at the time. By contrast, those 'who chose to sin on this festive day survived; the town brothels on the eastern outskirts of the city were spared' (2000, 75). The pretensions of Leibnizian theodicy were instantly shattered, thus undermining a key source of intelligibility in the natural world. Previously, it had been reassuring to understand natural disasters morally, as part of the best possible divine plan. By helping to remove purpose from nature—or simply rendering this sort of explanation absurd—the earthquake thus inaugurates the separation of natural and moral evil, which, according to Susan Neiman, is 'part of the meaning of the modern' (2002, 5).

Although there was scapegoating aplenty on account of the earthquake, just think how much worse it could have been had appeal to divine punishment for sin been the uncontested explanatory norm then. On the other hand, had leaders and subjects in the seventeenth century faced the challenges of climate change in the following century—after Lisbon and *Candide*—it is not too far-fetched to suppose they might have acted less objectionably on the whole. The response to the Lisbon earthquake expressed a general desire to figure out where agency, divine or human, was operative and where it was not. In domains where it was not, purely naturalistic explanations would suffice and, in the best-case scenario, even allow us to control future events in those domains. Explanation and the quest for control gradually replace the urge to punish. For example, partially in response to Lisbon, the science of seismology began its long development, its goal being both to understand how this fully natural phenomenon works and to protect human populations from earthquakes by setting up observatories, relocating vulnerable people, redesigning buildings, and so on.

The phenomenon of anthropogenic climate change is, in a sense, the next chapter in this story. Here we have a large-scale geophysical disruption, one underway already but building significant momentum such that its most dramatic effects will not be felt for a generation or two (and beyond). This episode of climate change is unlike previous ones chiefly because we know quite a bit about both its causes and its likely effects. And yet in spite of this knowledge, our failure (so far) to address the problem is, to say the least, deeply worrying. The global average temperature anomaly during the Little Ice Age was merely –0.6°C relative to the period 1000–2000, whereas (as I show in Chapter 3) we are currently on course to hit an anomaly of almost +5°C by 2100 relative to the pre-industrial baseline. How hard societies are hit by climate change is a function of both the size of the anomaly (as well as how abruptly it arrives) and their degree of preparedness. With respect to the general social uptake of the relevant facts—as opposed to what most scientists see clearly—we are both ignorant that we are drifting into uncharted climatic territory and unprepared for the threats awaiting us there. Because of this, and using the seventeenth-century as an analogue, it certainly looks as though we are on the verge of allowing a human catastrophe of untold proportions to unfold over the next century and beyond.

It is quite remarkable that more than 250 years after Lisbon we cannot manage to act meaningfully to counter, avert, or even mitigate a disaster we can see coming from a mile away. As heirs of the Enlightenment project of structuring our worlds as much as possible in accordance with what reason uncovers, how do we explain this? Obviously, peccatogenetic explanations and appeals to punitive divine agency work hand in glove, and this way of seeing things is no longer available to us. And yet there is, I want to suggest, something not entirely amiss in this form of explanation. So long as we strip it of its theological connections, we can see that it sometimes makes a good deal of sense to explain a bad state of affairs in the world by pointing to something gone wrong in the hearts and minds of human agents. Two final points about the seventeenth century are salient. First, there is no reason to assume that the way large numbers of people responded to failed crops, floods, etc.—by turning violently on one another—was in all cases necessitated by trying circumstances. People were also agents in the spread of disorder. That remains relevant for our case. Second, unlike our forebears we are the principal causal source of the disrupted climate. If anything this second point, to the extent that we have not yet fully embraced it, makes us look even worse than our supposedly benighted ancestors. I have no interest in sin, but I do believe we must look to our disordered passions and beliefs to explain fully our failure on this front.

In the chapters to follow I am going to devote a great deal of time to the development of this claim by providing a virtue-focused climate ethics. And just as the diagnosis looks inward so too does the prescription. If we are going to find a morally defensible path through the climate crisis we need to become better people, and that means cultivating the virtues—most pertinently justice, truthfulness, and rational hope. But this raises two large questions. First, why are these three virtues being picked out for special philosophical treatment? Second, isn't there something objectionably private about the cultivation of the virtues that makes it a retreat from the world's challenges?

1.3 Three Virtues of the Anthropocene

The quick answer to the first question is that justice, truthfulness, and hope are virtues expressing principles or ideals or values that many of us already espouse. But that answer forces me to expand on the nature of this 'we'. This is an especially important question with respect to the climate crisis, for there are at least three large groups whose interests are at stake in what we do (or fail to do) about the problem and whose voices thereby demand some degree of recognition: 'the global prosperous' (I'll define this group just below), the global poor, and future generations. In this book, I address the ethically unique situation of all three, but my focus is overwhelmingly on the moral failure of the global prosperous in attending to the vital interests of people of the future. Does this leave the global poor out in the philosophical cold? Not really, because while I maintain that our primary ethical objective at the moment ought to be to curb consumption among the global prosperous in the interests of avoiding future disaster, I also agree with those who claim that the lion's share of whatever remains in the 'carbon budget' belongs, by right, to the global poor. I present no extended argument for the second part of this claim, however, because it has already been ably defended, most prominently by Henry Shue (2014).[2]

Instead, I am quite consciously directing my critique and analysis at a specific subset of current humanity. To some extent, beyond calling it a subset, I have no precise idea of its scope. But I'm not sure this is a problem. Consider what David Lewis has to say on the topic applied to a different problem:

I say X is a value; I mean that all mankind are disposed to value X; or anyway all nowadays are; or any ways all nowadays are except maybe some peculiar people on distant islands; . . . or anyway you and I, talking here and now, are; or anyway I am. How much am I claiming?—as much as I can get away with. If my stronger claims were proven false . . . I still mean to stand by the weaker ones. So long as I'm not challenged, there's no need to back down in advance; and there's no need to decide how far I'd back down if pressed. (2000, 85)

The values, ideals, and principles I have in mind have to do with the moral equality of all humans, the pride of place we give to scientific descriptions of the natural world, and the hope we cherish for a better future. I certainly do not want to claim that these are actually endorsed by 'all mankind'. But when set out this starkly their power and broad appeal are surely difficult to gainsay, so I'm also confident that what follows is not just a conversation between 'you and I' (or worse yet, with myself alone).

An important target of my critique are what Chris Hedges calls the 'liberal class' and the institutions it has created: the progressive churches, the press, the university, etc. Members of this group are, for the most part, liberal, tolerant, progressive, prosperous, and educated. According to Hedges, the liberal class has been 'seduced' by a false idea of progress which it believes is 'attained through technology and the amassing of national wealth, material goods and comforts' (2010, 102). Their complicity in the ongoing degradation of the natural environment and the rise of the global plutocracy is fuelled by 'a craven careerism and desire for prestige and comfort' (Hedges 2010, 103). For the most part I agree with this, but as regards climate change inertia we must cast our net wider. My broader target is therefore what I will refer to throughout as the global prosperous, which I'm stipulating to cover those who (1) are on the upper end of the consumption spectrum relative to the rest of the world;[3] (2) espouse broadly liberal-democratic moral and political ideals; and (3) value scientific rationality.

I don't think you need count yourself a member of Hedge's liberal class to fit this description, at least if we understand the liberal class to refer to those Americans who self-identify as 'liberals' (as well as this group's analogue in other countries). For example, although it surely excludes far-Right conservatives (to whom (2) and/or (3) are inapplicable, though lots of them are rich), it is meant to apply to many more moderate social and political conservatives, and not just in the U.S. There's no obvious inconsistency in a relatively affluent person endorsing the values of democracy and scientific rationality on the one hand and locating herself on the non-extreme Right of the political spectrum on the other. Such a person might think that there is nothing at all progressive or liberal about her political and moral values. Nonetheless, my argument will be that this person and everyone else in this very large group have overriding moral reasons to treat people of the future as genuine moral subjects. This entails constraining present pursuits in specific ways so that the interests of those people are taken account of. But despite this broadening exercise I recognize that there will be types who fall outside the scope of the analysis: the hardcore climate denier who really does not see a problem, the millenarian religious enthusiast pining for the end times, the wealthy generational chauvinist who is sincerely unimpressed by the suggestion that he and his ilk are consuming the future's material capital at breakneck speed. And so on.

There are three things to say about this challenge. First, I do not neglect these agents altogether. Indeed, since the pathologies come out most worryingly in certain forms of climate change denial, I will have a good deal to say about them in the opening sections of Chapter 5. But, again, they are not the focus of my analysis. Second, nothing I say here precludes a fuller analysis of the agents just mentioned along virtue-theoretic lines. Indeed, I think my analysis invites just such a treatment, though I do not offer it myself. Finally, no single critique can do full justice to the complexities of our collective inertia on this issue. In the end, I'm inclined to revert to Lewis's apology. There's no need for me to 'back down in advance' just because my account doesn't fit *all* the relevant players.

But more can be said. If the global prosperous *did* radicalize its existing moral commitments in the manner I lay out, we would be substantially further along in solving the problem. In this case the pathological would simply be left in the cultural dust, where the morally incorrigible belong. My group, by contrast, is by nature corrigible, so it also makes pragmatic sense to focus on them. It is true that climate change involves or implicates all of us in one way or another: the amorally unmoved, the simply unaware, the global poor, the liberal class, the 1 per cent, and more. We are interlocking pieces in a global economic system which seems, now, to be largely self-sustaining. However, as Donella Meadows points out, if we want to change or slow the destructive patterns of such 'world systems' we need to find 'leverage points' within them. And we should seek these points chiefly in ' "success to the successful" loops, any place where the more you have of something the more you have the possibility of having more' (Meadows 2009, 156).

How does this work in the case of the global prosperous and climate change? Many of us want to be the sorts of people who embody and strive for justice and truthfulness. We want these values to inform our motivations, our emotions, our beliefs, and our actions. But they don't, at least not fully. Much of my analysis—Chapters 4 and 5—therefore focuses on the moral weakness and self-deception of the global prosperous. We don't act consistently on principles we otherwise endorse because, seduced by consumption, we lack full self-control; and the full truth about climate change makes us anxious, so we find ways to flee, distort, or conceal it. These forms of motivated irrationality prevent us from being, respectively, fully just and truthful people. However, the desire of the global prosperous is a powerful potential force for change.

Through relatively unconstrained consumption this force is currently exacerbating the problem of climate change. The economic behaviour of this group is thus a clear example of a destructive success to the successful loop. Because of the access to democratic institutions much (though not all) of this group enjoys it can use this very force to alter the global system. I show that there is no other way to do this than through significant desire-constraint on its part, but my claim is that members of this group can, in principle, be persuaded to do this by their own moral lights. This makes them an ideal leverage point in the world system.

This brings me to the second question about the three virtues. While discussing Parker's analysis of seventeenth-century responses to abrupt climate change I noted that I did not want to be misconstrued about the sort of inward-looking move I was advocating. Sin is no longer in our repertoire of moral concepts. But I also think it would be deeply misguided at this moment in our cultural history to take Candide's advice and cultivate our gardens, if this is taken to advocate retreat to the private sphere. The salient feature of justice, truthfulness, and hope is that each is an essentially outward-looking or public virtue. They are virtues of connection, each in a unique way. I think of the virtues generally in a manner that stresses their reason-responsiveness. They are, among other things, ways of seeing the world, and it follows that there is something out there for them to be right (or wrong) about. Moreover, once we grasp the sense in which each virtue is reason-responsive we can see clearly what its opposing vices look like as well as what subsidiary virtues and emotions it requires to do its work well. Rather than trying to defend this idea in general terms it might be more fruitful to see how it can explain the way our three virtues typically function.

Justice and hope are perhaps the easier cases, so I'll deal with them first. Plato was the first to see that justice in the soul and justice in the city are correlative states of affairs. Justice is thus a form of *social* outwardness. It looks specifically to other people, their vital interests, and the institutions constructed and maintained to safeguard those interests. More particularly, justice has to do with who gets what in social distributive systems, and this is why the vice of greed is the key threat to justice. But justice considered as the disposition to distribute social benefits and burdens fairly cannot do its work without help. I will argue that it needs the assistance of courage, anger, shame, and honour. Hope is a form of *temporal* outwardness. It is directed at the future but in a way that is constitutively lucid and active. That is, it is opposed to (and by) the vices of wishful thinking and cynical or apathetic retreat. It neither sugar-coats dire probabilities nor wilts in the face of them. But it cannot do its work without imagination, moral seriousness, humility, and courage.

The same structure applies to truthfulness, but here things are more complex. This is because with respect to the climate crisis we need true beliefs about both the natural world and whatever settled moral convictions allow us to respond to the crisis appropriately (for example, those concerning the fair allocation of costs for mitigation and adaptation). This means that the virtue of truthfulness points outward in two distinct ways: towards the *ecological*, understood so as to encompass facts about earth's biogeophysical systems; and towards the *moral*, specifically a version of the 'polluter-pays principle'. So denial of truth is the vice opposed to truthfulness, but with respect to climate denial this has many faces, partly as a result of the two distinct objects of the virtue. As with justice, truthfulness requires courage, but also autonomy, humility, and conscientiousness.

Setting out assiduously to cultivate justice, truthfulness, and hope will not therefore result in the objectionable privatization of the problem because these virtues turn us towards reasons emanating from the world. Finally, because of the specific ways in which they force us to deal with the world, I have called justice, truthfulness, and hope virtues of the Anthropocene. As I explain in detail in Chapter 2, some geologists have taken to dubbing our age the Anthropocene because, in contrast to previous epochs, and chiefly because of our awesome technological reach, we are now collectively exercising significant causal power with respect to the earth's macro-systems (its great cycles, for instance: carbon, phosphorus, nitrogen, water). What we do to and with these systems is, whether we like it or not, going to affect the availability to future generations of material resources in a way that has inescapable moral implications.

The dream of the Enlightenment had to do with the expansion of control over nature and society. Those thinkers took this to be a project and process rooted in both our newfound causal powers—that is, our expanding technological reach—and the rationality of our methods of inquiry. But one of the key differences between us and the eighteenth-century Enlightenment is that we now see that causal power over nature does not entail complete control of it. We can intervene like never before in natural systems, but we do not have full control over the effects of our interventions. This is because along with the discovery of our causal powers we have also learned that the systems we affect respond to our interventions in non-linear and therefore unpredictable ways, and that these responses will in turn affect human systems, also (sometimes) in non-linear ways. We are now stuck in a feedback loop between natural and conventional systems that is both unprecedented for our species and, I believe, will ultimately undermine the very distinction between the natural and the conventional. We need to learn how to manage this state of affairs wisely.

There is no way to reduce our challenges to a formula, no method that will allow us to engineer our way to full predictability and safety. For the foreseeable future our lives will be marked by increased risk and moral danger. This is the new and inescapable context of our ethical choices in the Anthropocene, and part of our job now is the epistemic one of describing it correctly. As I have just hinted, what we really need is wisdom, but since this is probably an irremediably fuzzy concept my three virtues are together meant to be a good enough proxy for it. The world will be significantly altered by our collectively applied technologies in the decades and centuries to come. The focus on justice, truthfulness, and hope is meant to reflect that fact and give us a morally decent way to negotiate the complexities it will involve.

1.4 Collective Action and the Situationist Challenge

Still, it might be asked why we should adopt a virtue-ethical framework to the problem of climate change in the first place, especially when there are compelling normative-ethical competitors in the field. The standard move at this point would be for me to show how it is that deontology, contractualism, and utilitarianism fail to describe the problem as well as virtue ethics, or how their prescriptions are less far-reaching or intuitively evident, etc. That sort of approach can be fruitful, but it demands a lengthier, and inevitably less focused analysis than the one offered here. Instead, in the rest of this chapter I will motivate my own approach in two ways which, taken together, should go some distance towards vindicating it. First, I will argue that climate change is, as many have noticed, a huge collective action problem and that a focus on the virtues provides a uniquely compelling way of looking at such problems. Second, I will engage the situationist critique of virtue ethics, showing that a (perhaps ironic) upshot of this school of thought is that we need robust virtues, especially in a time of crisis like the one we face. The committed utilitarian or deontologist will doubtless point out that I have begged some important questions along the way, but I hope that the conclusions I offer will be compelling enough to provide a measure of compensation for this fault.

So far we have no sustained, book-length treatment of the virtues applied to the climate crisis. This is somewhat odd, in view of both the increasing amount of attention virtue ethics has been receiving in the last thirty-five years or so from philosophers and the very recent rise of environmental virtue ethics. But many philosophers have come to realize that a virtue-oriented approach to climate change might be fruitful. The reason for this is that climate change is, among other things, a huge collective action problem. It is the product of a large number of people engaging in actions which, taken alone, are inconsequential but which are extremely harm-causing in the aggregate. Each participant—individuals, firms, nations—calculates that it will lose competitive advantage if it constrains its pursuit of the offending activities (unless everyone else does so as well), the predictable result of which is near-universal non-constraint leading to suboptimal outcomes.

I don't make these games the focus of my analysis, mostly because describing complex social issues through the lens of game theory strikes me as artificial and potentially distorting. But the idea does come up occasionally (especially in Chapter 4) so it might be useful to distinguish the forms these structures take in practice. For example, I think it is useful to characterize the behaviour of nations in the ongoing struggle to come up with meaningful and binding emissions reduction targets as a tragedy of the commons. The atmosphere, forests, and oceans are carbon sinks, with a finite capacity to absorb our emissions if we want to avoid dangerous interference in earth's macro-systems. But since emissions are tied to economic growth, each nation calculates that it is better off with one more year or quarter (or other fixed period) of emissions-based economic activity. This is like the herdsman adding one more cow at the economic margin. Even if it represents a cost to the collective using the atmospheric and oceanic commons—polluters need not deny the relevant science outright—the cost is spread throughout that collective while the growth-gain accrues entirely to the emitter.

On the other hand, the sort of status-based competitive consumption engaged in by ordinary individuals—especially, but not only, the global prosperous—is probably best characterized as a kind of prisoner's dilemma. Dale Jamieson has argued that the key feature of these collective action problems (though the point extends to other forms as well) has to do with the way the players key their behaviours to perceptions about how others are going to act. That is, we make our behaviour 'contingent' on what we believe about others' likely behaviour, and this is what causes the suboptimal outcome. The solution, says Jamieson, is to look to alternative 'generators of behaviour':

> Instead of looking to moral mathematics for practical solutions to large-scale collective action problems, we should focus instead on non-calculative generators of behaviour: character traits, dispositions, emotions...When faced with global environmental change, our general policy should be to try to reduce our contribution regardless of the behaviour of others, and we are more likely to succeed in doing this by developing and inculcating the right virtues than by improving our calculative abilities. (2007, 167)

Echoing Jamieson, Ronald Sandler argues that it is a condition of adequacy for a normative theory capable of addressing large-scale collective action problems like climate change that its 'evaluations of an agent's actions [are] not overly contingent on the actions of others' (2010, 172).[4] Both philosophers conclude that we need to look to the virtues as a way out of these problems. In saying we need to be 'virtue-theorists' Jamieson is not claiming that we should abandon appeal to consequences as a criterion of right action. He is asking consequentialists only to become virtue-theorists, not to abandon consequentialism. I will return in a moment to the claim that turning to the virtues will help us find a way out of collective action problems. For now, following up this reference to consequentialism, I want to be more specific about the type of virtue ethics I will be leaning on in this analysis.

Slote's notion of an 'agent-focused' virtue ethics is helpful here. Agent-focusing is teleological (or consequentialist in Jamieson's formulation): the rightness of virtuous dispositions is defined by the ends they help bring about.[5] These of course can vary. Aristotle thinks that the virtues are teleologically aimed at our flourishing and, broadly, this is the view I will adopt in what follows.[6] However, instead of laying out the merits of this approach abstractly, in subsequent chapters I will describe the target virtues in detail—including their connections to our motivations, emotions, and actions—and let readers decide for themselves whether or not cultivating these virtues can help us solve real moral problems like climate change. The ideal upshot of this exercise is that the morally desirable consequences associated with cultivation of the virtues—in our case, averting some of the worst outcomes associated with runaway climate change—will become perspicuous. If this happens, we will by that fact have reason to make virtue-focused assessments of agents primary, at least in the limited context of the exercise. I will add three further points about an approach like mine that emphasizes the connection between the virtues and human flourishing.

First, I assume the correctness of a (qualified) naturalized account of what flourishing consists in. I don't mean that we should look to, say, sociobiology to explain morality for us. We are, alas, quite a bit more complicated than ants, whatever E.O. Wilson implies to the contrary. Most generally, the philosophical naturalist claims that we humans are, though rational, also material, finite, and dependent beings and that whatever we say about the good life should be consistent with these basic features of our make up. But this is a fairly weak explanatory constraint, one that is, for example, compatible with the claim that we are also free in a meaningful sense (a claim that is important to my purposes in this book). More particularly, the commitment to naturalism means that we should (a) eschew all supernatural explanations of our nature, our duties, and the good life; and (b) appeal to scientific explanations, both natural and social-scientific, where they are relevant to our normative concerns. For instance, if we expect to say anything meaningful about how humans might flourish in, say, fifty years, we require quite a lot of data—concerning mean temperatures, levels of ocean acidity, population levels, the energy mix, and much more—about what the world will look like on various climate paths. That information should come from a diversity of sources, most of which are purely empirical in orientation. This information can tell us whether or not the material preconditions for human flourishing will be available to future people (or to what degree they will be available). But this concerns only the material preconditions for flourishing. Since the concept of flourishing is intrinsically normative that is as much enlightenment as we should expect from science about it.[7]

Second, while the virtues are necessary for our flourishing, I agree with Aristotle that they are not sufficient and that luck—for our purposes, specifically the sort that provides us with relative climatic stability—is also required. I say that climatic stability is largely a matter of luck because, although we are now in a position to do something about it, the condition in which we ultimately leave the climate is going to be a matter of good or bad luck for generations to come. One way or another this will affect the ability of future people to flourish. Nor, third, does a focus on the virtues force us to abandon altogether deontic language in our moral assessments. In Chapter 2, for example, I talk about the duties of extended cosmopolitanism. Sometimes it can look as though these duties are independent of our dispositions, that they are the directives of pure practical reason, but that is an illusion. Since Kant and his followers are the main historical representatives of modern moral cosmopolitanism, deontic language comes naturally in discussing this concept, but, as the rest of the book makes clear, to be effective I believe that cosmopolitan principles must be firmly rooted in certain dispositions (and that the latter are themselves complex amalgams of emotions, beliefs, intentions, tendencies-to-act, etc.).

Coming back to Jamieson, is he correct in arguing that the virtues will help us avoid the pitfalls of collective action problems? I think so. The key to his account is the idea that the virtues impart a measure of firmness—non-contingency—to our characters that allows us to make our decisions non-reactively. Instead of waiting to see what the other players are going to do, I assess the situation's demands on me *qua* just, compassionate, hopeful, etc. person. This is a crucial point. Collective action problems typically result in suboptimal outcomes because of the reactive nature of our decisions in them. Typically we adopt a wait-and-see approach to decisions in contexts like this, or we simply extrapolate from our knowledge of various players' past performances in relevantly similar situations. There's nothing wrong in principle with being sensitive to the designs of others with whom one is forced to think and act, but when every agent bases her decisions mainly on predictions of this sort, everyone loses. Fear of being a sucker and the desire to get ahead become our dominating motivational impulses.

Think about collective action problems rooted in competitive consumption. Suppose I'm contemplating reducing how much I consume of a certain product, one people like me take pride in being seen to consume. I pick this good because the evidence suggests that if a critical mass of people refrain from consuming it, meaningful reductions in greenhouse gases will follow. The complex but corrosive thought that so often follows is that my action won't make a difference unless enough other people act the same way. This is sometimes referred to as the 'negligibility thesis', according to which my contribution to a collectively realizable outcome is inconsequential.[8] If I believe this, and have in addition no reason to think that others will do their part to bring about the outcome, then I likely won't be motivated either to engage in actions that I believe are otherwise worthwhile or to refrain from actions I believe are otherwise problematic. If others refuse to reduce while I do so, I'm just a sucker. I lose that source of pride, with no pay off. So I don't reduce. I don't think there's any way to appreciate what is wrong with this reasoning within the confines of game theory. But we do need to underscore what disastrous outcomes it is bringing about for the global climate.

Game theory is not the only theoretical culprit here. The most important use we can make of the moral imagination right now is to think about what the world might look like in roughly one hundred years if we do not change course rapidly. With everything climate science has revealed to us about the possibilities, this should not be too difficult to do. However, according to evolutionary psychology and sociobiology it is not in our nature to take the long view climate ethicists demand of us. We evolved to rise only to more immediate challenges and our hunter-gatherer brains are simply not wired to think about, let alone be motivated to avert, negative consequences as far-flung in the future and as unconnected to kith and kin as these ones are. In assessing this claim, let's begin by looking at how we might react to morally loaded stories set (roughly) one hundred years in the past. For example, Pat Barker's stunning multi-novel reconstruction of the First World War years—the artistic pinnacle of which is the depiction of soldier and poet Siegfried Sassoon in *Regeneration*—opens up an entire moral world for us. So does Solzhenitsyn's *Gulag Archipelago*. These stories don't involve us in a way that evolutionary psychology and sociobiology can make much sense of and yet they clearly motivate us to think, feel, and act differently than we had been before encountering them.

Similarly, we patently *can* think about and be changed by contemplation of morally significant far future events. It takes a little more cognitive work and moral imagination than is involved in our reaction to more immediate threats—a looming economic recession, that figure huddled in the shadows—but how else can we explain our perennial fascination with fictional depictions of far future calamity? They need raise no pressing existential issues for us and yet films and novels like *Never Let Me Go*, *The Road*, or *Planet of the Apes* have the power to disturb us, to make us question ourselves because of the harms our actions are depicted as bringing to people who are currently non-existent and who will (for the most part) be genetically unrelated to us when they come into existence. This is a purely moral reaction, aimed in the first place at their plight not our own. We would not experience it absent some capacity for concern about the long-term prosperity of the species. That a contemplated scene of human suffering is temporally remote from us, in one direction or the other, has little to do with its power to stir the moral imagination. Evolutionary psychology and sociobiology are blunt explanatory tools. But since they also function to reinforce our sense of helplessness, it is not at all surprising to see such widespread appeal to them in the face of the difficult moral truths and challenges of climate change.

We therefore need to ask some hard questions about where we are headed and who we are. Will we wreck the global climate utterly because we are anxious about losing a bit of competitive ground to our neighbours? Will we allow greed, harnessed to help us secure more and more of our culture's endless supply of frivolities, to plunge us into a new age of hyper-violence and communal fragmentation? We hear a lot about the fog we seem to be in about climate change, but can it really be the case that we will sleepwalk into the sort of worlds portrayed in our most lurid dystopian fantasies? Are we going to watch passively while half or more of all non-human species perish forever because we refuse to relinquish an obviously perverse and totally dispensable mode of practical rationality?

If the answer to these questions is yes (as it appears to be) what should we think of ourselves? Are we devoid of moral imagination or are we hopelessly depraved? If we are so incapable or unwilling to try a bit harder to save civilization, some good old-fashioned Nietzschean self-loathing might be in order. It is no abrogation of philosophical naturalism to claim that we are free to reject the demands of purely instrumental rationality as set out by decision theory, free to reject evolutionary psychology's pinched vision of the moral imagination, and free to redirect our moral powers to a frank assessment of what morality demands of us as members of an intergenerationally spread community of equals.[9] Nor is there anything necessarily masochistic about the sort of negative self-assessment I am urging.[10] Sometimes such thinking can reveal options covered over by complacency. Nietzsche thought it was a problem for the future of the species that a certain type of prosperous and comfortably liberal-minded citizen—what he calls 'the last man'—was incapable of despising himself. He, Nietzsche, knew that material comfort and political power make people stupid by shutting down their capacity to imagine a better future. This is one reason he loathed the sentimental and chauvinistic culture of Bismarckian Germany.

I wouldn't have written this book if I thought this was the last word about us, but I do think a dose of self-directed disgust would be therapeutic at the moment. Beyond that, in the face of this massive crisis it is imperative that we identify the settled dispositions that can guide us in building a collective bond with future generations and then dig in our heels as we make choices in concert with other members of the human community. Our decisions must be framed by consideration of the sorts of people we want to be, not how we believe other agents' choices and actions will affect our material or reputational standing in the collective. This involves cultivating the ability to react to the latter considerations—in ourselves, our friends, our political leaders, etc.—with the contempt and disgust they deserve. There is no guarantee that this stance will save the day, but if widely adopted it might. At the very least it will clean our hands and impart a measure of moral integrity to our actions. This can all sound self-righteous, but in the context of the moral calamity that is climate change is that really our most pressing worry? The studied avoidance of self-righteousness is often a high-minded disguise for moral superficiality.

Think of these claims and concerns, especially the role played by the concept of contingency in the forgoing discussion, in the context of the recent rise of situationism in social psychology and philosophy. Inspired by the obedience experiments performed by Milgram and Zimbardo, philosophical situationists like Doris and Harman argue that cross-situational stability of character is a fiction, and that this calls into question the theoretical foundation of virtue ethics because the attribution to agents of robust or global character traits is an error (Harman 1999, 2000; Doris 2005).[11] Rather than being the product of enduring dispositions, these researchers suggest that our behaviour is significantly determined by the variables of the situations in which we find ourselves. The situationist attack on the virtues is complex, but one of its key claims (for our purposes) is that agents do not display and should not aspire to display substantial trait consistency across diverse situations. Here, our choices are, and should be, *contingent on* beliefs about our circumstances, including, presumably, the imagined or predicted decisions of other agents.

The claim is both descriptive and normative but is dubious on both counts. First, Nancy Snow and others (including, lately, many social psychologists themselves) have shown that the descriptive claim trades on an objective conception of situational sameness or difference. Here, it is true that cross-situational consistency is rare. But when it comes to agents' subjective apprehension of situations, things are different. Agents in fact display a great deal of cross-situational consistency in their behaviour when their own interpretations of situations are emphasized (Snow 2010; Sreenivasan 2002). As for the normative claim, Doris, for example, thinks we can still talk about the traits we ought to cultivate. However, this language is proper only for 'narrow' or 'local' traits, those which are confined very tightly to context. If one is especially temperate in the gustatory domain, for instance, then one would do best to continue to be temperate there consistently, refining one's ability to avoid contrary-to-virtue temptations as they arise. One should not, however, expect to find new domains for this trait's application, the sexual domain for instance. Indeed, the idea seems to be that most of our moral failures occur when we attempt to expand our moral vision beyond these contexts.

Take courage. For the situationist, courage, like any other trait, will have many faces, all of them highly attuned to context. For example, we should not expect the person of physical courage to be morally courageous, nor should we expect moral courage to be consistently displayed towards diverse objects of moral concern. Courage in the face of a police interrogator's threats does not, and should not reasonably be expected to, predict for courage in the face of a spouse's overweening demands (Doris 2005, 64–5). Nor, presumably, is there any normative connection between acting courageously to protect the interests of those near and dear and doing so on behalf of members of future generations.

Montaigne captures the alternative position nicely:

> A man who is truly brave will always be brave on all occasions . . . If he cannot bear slander but is resolute in poverty; if he cannot bear a barber-surgeon's lancet but is unyielding against the swords of his adversaries, then it is not the man who deserves praise but the deed. (1991, 378)

Doris argues that an ideal like Montaigne's is pathological, and that we should instead opt for 'substantial behavioural inconsistency' which although it 'may confound our interpretive and ethical categories . . . may also signal sound mental health' (2002, 65). Which normative ideal is better at signalling sound mental health? Consider Tim O'Brien, a conscripted infantryman who served in Vietnam 1969–70. He relates the story of a girl in his fourth-grade class who was dying of cancer. She wore a headscarf to hide her balding head and was ridiculed for this by her classmates. Says O'Brien: 'I should've stepped in: fourth grade is no excuse. Besides it doesn't get easier with time, and twelve years later, when Vietnam presented much harder choices, some practice at being brave might've helped a little' (Quoted in Miller 2000, 64). O'Brien's report says something important about cross-situational consistency of character. It shows us an agent who has made substantial moral progress by recognizing that courage in one sphere can serve as practice for another. Further, although O'Brien is talking here about moral courage, he might just as plausibly have been talking about some other virtue—benevolence, justice, or compassion for example. As I have said, the situationist cannot recognize O'Brien's report, to the extent that it expresses an ideal of cross-situational stability, to be anything more than a kind of pathology. That, however, is patently false to our sensibilities in cases like this. When Doris says that the situationist ideal 'confounds our moral and interpretive categories', this is the sort of upheaval we should expect. At the very least, we should be wary of such counter-intuitive reinterpretations of our practical lives.

Moreover, when times are tough we will likely have cause to worry about the stability of even our local traits over time, and this means there is a good deal of artificiality in the distinction between local and broad traits. One way to put this point is to say that when crises arrive, they challenge us by altering the situations we are in. Suppose I've got a trait that is locally specifiable as 'loyalty to close friends'. Previously, this virtue had been exercised in a variety of related circumstances with my close friends: I helped Dave move though I didn't really feel like doing so, I took time away from my own projects to drive Frank's kid to her soccer game when he was indisposed, I took Maria out for beers and listened to her grouse about her boss and her marriage, and so on. But what happens when my loyalty to any of these people is truly tested? What if Dave can't find a job and needs to move in with my family, or the stock market collapses and Frank and I are now competing for this or that bundle of scarce resources (like the *same* job), or (being a committed environmental activist) Maria is arrested for monkey wrenching an oil pipeline and names me as an accomplice (we *had* discussed these things, after all, though I had no idea how serious her intentions were)? How might the situationist notion that we should cultivate only local traits work at explaining my new moral challenges?

Naturally, I will understand the virtue's field of application by reference to my past relations with Dave, Frank, Maria, and other close friends. However, in the imagined cases *the situation* has now changed so much that I am either forced to find new ways to apply the old virtue or give up altogether on the project of being loyal to my close friends since the local trait now appears to be a useless moral guide. This is because loyalty now appears to demand a level of sacrifice and risk on my part whose possibility I had never before contemplated. It is one thing to listen sympathetically to Maria's complaints, quite another to, say, lie under oath about our past discussions (as she has asked me to do). If I decide that I need to stick by my friends were any of these more serious things to happen, I need a stable disposition, and this amounts to expanding the virtue's field of application to cover hitherto unforeseen events or challenges. In other words, local traits will be temporally stable only if there is very little alteration in the circumstances of their application over time. But as Snow points out, such alteration occurs quite frequently:

> We need to cultivate our inner states as indemnity against the day when our social sustenance is taken from us—which is not, unfortunately always statistically rare. When our marriage breaks up, when our loved ones die, when our mortgage is foreclosed in a housing crisis, when a hurricane or a tornado or a stock market plunge wipes us out, we need the personal wherewithal to pull through, despite the demise of the social supports that once sustained us. (2010, 7)

The virtues can help 'indemnify' us against misfortune in these cases, but only if we constantly work at testing them in new circumstances, building up their stability for future crises. This is a large part of what moral education consists in.

It doesn't much matter whether we label this process broadening the original virtue or stabilizing it so that it holds firm in trying circumstances. The ideal effect is the same either way. Since climate change is and will increasingly become the ultimate situation-breaker, this is especially pertinent to a virtue-focused climate ethics. If all of this is right, the situationist's normative claim is overstated and he cannot respond by pointing to the descriptive claim—that there are in fact no cross-situational dispositions—because this too, as we have seen, is false. Situationism is a deeply valuable contribution to our understanding of how situational pressures can threaten us morally. It counsels us to be wary of over-reliance on the traits we actually have and it pulls us out of our naiveté about our characters. It should force us to strengthen the external 'scaffolding' of our lives, the social and institutional structures on which the successful moral life depends.[12]

The irony of the situationist challenge to virtue ethics is that it has shown us how much we need the virtues, especially where contrary-to-virtue pressures become intense and even where the relevant social scaffolding appears to be sound. In the crises we are likely to face this is a crucial insight. As we will see, there is a non-negligible chance that we will see a marked rise in inter-group violence and crimes of atrocity as people around the globe face severe climate change–induced deprivations. Situationism has shown us how social catastrophes like this can develop and spread. It has revealed the quotidian face of evil, showing us in the process how important all that fallible social scaffolding can be. We need the scaffolding, but no more than we need the stable virtues of justice, truthfulness, and hope and their moral psychological allies. Together, these resources will be our best indemnity against moral and social chaos in the coming decades and centuries. I am not convinced that bare appeal to the rights of future people or to abstract moral principles—grounded in welfare maximization, the principle of universalizability, or anything else—will do the job here. There's nothing a priori about this scepticism, it's just that the peculiar mix of cynicism, self-deception, weakness, indifference, and psychic inertia that currently has us in its grip demands an ethical perspective that is more psychologically nuanced than these approaches tend to be.

Every catalogue of the virtues expresses a historically particular conception of the good life. The dispositions it demands are intimately related to the moral and political values and ideals of some time and place. Most of these catalogues are culturally complex, the partly accidental and partly conscious product of social hybridizing, historical amnesia, and sometimes painstaking conceptual accretion. This is certainly true in our case. Although our understanding of the virtues and vices is rooted mainly in ancient Greek culture, we are also children of the Enlightenment. Though some have suggested otherwise, the notion that we can somehow shrug off this latter part of our heritage altogether—as opposed to fitting it to our historically particular needs—strikes me as pure fantasy. But in contrast to its ancient forebear, modern morality lacks psychological depth.[13] So why not try and understand our most pressing moral problems—and by extension ourselves—with a virtue ethics inspired by Enlightenment values and ideals? For the most part, philosophers have treated these two ethical streams as separate and mutually antagonistic. The Enlightenment is universalizing while all virtues ethics is more or less parochial (not necessarily in a pejorative sense). When it comes to our most fundamental moral and political group allegiances we must choose between the polis and the cosmos. For reasons that will be laid out in the chapters to follow, I reject this dichotomy.

1.5 Conclusion

I conclude with an outline of the book's main arguments. Having invoked the Enlightenment in this chapter, the task of Chapter 2 is to lay out what I'm calling the Anthropocene Project. This is explicitly an extension of the older Enlightenment project, with its focus on moving history forward by expanding the reach of science, the quest for justice, and the hope that we can make the world better in the future. The peculiar contribution to this story made by the concept of the Anthropocene has two key features. The first is that the cosmopolitan ideal—the radical moral equality of all humans—that is so central to Enlightenment thought needs to be extended so as to encompass people of the future. The second is that we need to recognize both our new power to affect natural and conventional systems and our inability fully to control them. The two features come together in my argument. The lack of full control means that we will encounter surprises as we intervene in nature (as we must), and that these surprises might involve bringing harm to morally considerable things. We can nevertheless justify our interventions if and only if we are guided in the first place by cosmopolitan ethical ideals (again, extended to take in future generations). I articulate all of this as a new way of understanding what we mean by 'moral progress', again a quintessentially Enlightenment concept that we have good reason to make our own.

If the cosmopolitan ideal is essentially integrative then the most pressing threat to it is the potential fragmentation of the human community and the inter-group violence this tends to breed. In Chapter 3 I show that there is good evidence to suggest that the climate path we are on is taking us in this direction. If this is right, it means our civilization may be headed for the rocks in the decades to come, and this is a threat we must bear constantly in mind as we attempt to meet the challenges of climate change. Much of this chapter is empirical in orientation. Based on a careful analysis of climate science on the one hand and security and genocide studies applied to the problem of climate change on the other, I lay out the precise threat to global civilization posed by this phenomenon. I close with a consideration of the ways in which these new realities help to further motivate a virtue-focused approach to the phenomenon. Together, Chapters 1 and 2 define the specifically ethical features of the new age we are in, suggest how it is that all of this demands a new self-conception on the part of the species, and assess the threats posed to our future by a problem we are currently allowing to fester.

The next three chapters form the heart of the book. Each focuses on one of the three virtues whose joint cultivation can help us find a way to meet the challenges we face. The problem of climate change is extraordinarily complex—'super-wicked', as it's sometimes put—so any attempt to contain it is bound to be partial. My contribution is no exception to this rule, but I suggest that the best points of leverage for change within the global system are the already existing commitments to justice and scientific rationality of the global prosperous. In Chapters 4 and 5 I expand on this claim by uncovering why we have so far failed in this regard. My argument is that we can legitimately be charged with both moral weakness (Chapter 4) and self-deception (Chapter 5) in our response to climate change. In our refusal to curb significantly our emissions-based consumption we do not fully live up to our commitment to justice; and we flee, distort, or deny the facts about climate change—including, importantly, the polluter-pays principle—almost as soon as they are uncovered. But, paradoxically, this complex cluster of behaviours provides some cause for hope because the weak and self-deceived, by definition, recognize the authority of the moral and epistemic ideals they do not quite live up to.

So the next task, that of Chapter 6, is to explore how we might understand hope as a virtue for the climate crisis. I argue that the concrete though complex object of our hope should be that we will decarbonize our economy quickly enough to preserve a meaningful normative continuity between our generation and subsequent generations, where the connecting thread is the impartial sense of justice. Since it can be helpful to understand our hope against the backdrop of our fears, I also say something about what a future without this connecting thread might look like. I conclude the book, in Chapter 7, with an analysis of the phenomenon of forgiveness applied across the generations. I show that, despite the challenge of Parfit's non-identity problem, we may be *candidates* for the forgiveness of future generations, but only if we take meaningful steps toward the goal of decarbonization. This does not mean we will be judged forgivable, but even the judgement of unforgivability may be better than the only other alternative, that we will simply have been forgotten as the future descends into barbarism. That outcome is not inevitable, but all the evidence suggests that we must act quickly and decisively to avoid it.

2

The Anthropocene Project

2.1 Introduction

The Anthropocene is the age of inevitable human intervention in Earth's macro-systems. For as long as we remain on the planet—indeed, given the lifespan of carbon atoms in the atmosphere, possibly well after we're gone—our activities will affect these systems in significant and discernible ways. This will have profound impacts not only on us, but on virtually all other species as well. In the Anthropocene our reality-shaping collective agency and the radical openness of the future to that agency are emphasized. But, as we will see, there is a fundamental ambiguity here because our new causal power does not bring us full control over its products. To refer to the Anthropocene as a 'project'—the conceit I have adopted in this book—is obviously to go well beyond its specification as a geological epoch. To talk about the Eocene, Pleistocene, or Holocene as projects in the sense I intend would be absurd because no other epoch is defined by the (increasingly) consciously deployed and complexly ramifying causal power of a single species. Climate change is the most important phenomenon in the history of our species because it has revealed all of this to us for the first time and in the starkest possible way. The climate crisis forces us to ask fundamental normative questions about how we relate to each other and to the rest of the biosphere. We need to know why we have so far failed to address the problem and how we should move forward in a rationally justifiable way. This is the sphere of climate ethics.

This study begins by arguing that the most fruitful way of positioning ourselves in this new reality is to define the task of the Anthropocene as an extension of the older Enlightenment Project. We do not need to invent entirely new ways of thinking and being, because the normative conceptual materials we require are ready to hand.[1] But as we will see throughout the book, appropriating these materials thoroughly and honestly will have profound effects on the way we understand the human collective. I begin in this chapter with an elaboration of the idea of moral progress and show how it is linked to Enlightenment principles we should still accept. I also defend it against 'tragic' and pessimistic stances to our future. One of the signal achievements of the Anthropocene is our increased appreciation of the complex and interrelated systems—natural and social—in which our lives are embedded. We have learned that these systems function in ways that place limits on our ability fully to control them. Should we throw up our hands fatalistically in the face of this complexity or cling doggedly to an outdated dream of full control? To employ Giddens' terminology do we succumb to 'engulfment' or dream vainly of 'omnipotence'?

The next part of the chapter shows that we should stake out the ground *between* these two extremes, although this is a place of ineliminable risk. The argument here is that to manage this risk wisely all of our large-scale socio-technological and economic interventions must be actively subordinated to cosmopolitan moral principles, extended so as to take in the vital interests of future people. I support this line of thought by showing that generational membership is a morally arbitrary feature of agents and cannot therefore be used by present people to justify disadvantaging future generations. We have, it turns out, very strong duties of intergenerational justice. In preparation for the chapters to follow, I conclude with some thoughts on the abiding importance of a triad of virtues rooted firmly in Enlightenment soil: justice, truthfulness, and hope.

2.2 The Enlightenment and Moral Progress

The Anthropocene is above all a task, but how should we understand this task? Obviously we need to go beyond the claim that it is imperative for us to become more aware of the effects of our technological interventions into the natural and social worlds, though that is a good place to start. The next step is the really difficult one because having grasped the fact of our influence we must then realize that we have options among which we must choose. The task of the Anthropocene is most basically to recognize that the future is open to our collective agency in a way unprecedented for our species.

There are various ways to approach this. For example, to borrow a distinction from Daniel Innerarity, we can choose 'acceleration'—our current path—which is a faith in ever-expanding economic and technological progress. The idea is that we can solve whatever problems we will encounter in the future simply by ramping up the very processes that created the problems in the first place. This is what inspires claims like that of Exxon-Mobil's CEO Rex Tillerson that climate change is 'just an engineering problem' (Quoted in Daily 2012), or Bjørn Lomborg's claim that its negative effects can be fully countered by amassing more and more wealth (Lomborg 2010).[2] On the other hand, we can engage in largely private and reactive acts of 'deceleration'—survivalism, some forms of 'green consumerism', the call for 'Uncivilization' made by radically ecocentric groups like the Dark Mountain Project (2014) etc.—the goal of which is to retreat from accelerative processes judged to have gotten out of control (Innerarity 2012, 25–7).

Putting the task this way indicates that we seem to have difficulty getting beyond the ideological division between cornucopians (or Private) and neo-Malthusians to which we have been subjected for forty years or more. One side in this hoary contest clings dogmatically to what we now know to be a false principle—the idea that social progress can be achieved through the *unfettered* pursuit of technological innovation and economic growth—with the second then responding reactively to the perceived deleterious effects of this principle put into practice, a reaction that too often rejects modernity's laudable push toward a more integrated humanity. Sometimes the battle-ground is one or another version of the Enlightenment. At least implicitly—but sometimes out loud—accelerators take themselves to be children of that bold enterprise, while decelerators think the whole thing was a huge, hubristic mistake. Unfortunately, we seem to be shuttled back and forth between these two equally unpalatable options. As a result we are getting exactly nowhere in our efforts to solve large-scale collective action problems like climate change.

Still, it's not at all arbitrary that the Enlightenment should come to play such an important role in our thinking about the challenges we face. One reason for this is precisely the character of the Anthropocene itself. As I have already hinted, and *pace* the postmodernist aversion to grand narratives, the new epoch places us ineluctably as the protagonists of an historical narrative of undreamt-of scope. We are, like it or not, *responsible* for future generations because we are now shaping the material parameters of their lives like never before. The question then concerns what narrative and conceptual tools we have available to describe and respond intelligently to this state of affairs. And here it looks like we simply have not got anything nearly as well worked out, or as relevant, as the principles of the Enlightenment. If we eschew, as we should, both religiously inspired eschatologies and all forms of regressive neo-primitivism, we are left with little else to lean on.

But we should not despair of this because what we call 'the Enlightenment' is a deep pool, one brimming with ideas and principles to guide us through the crises we are going to encounter. Taking this notion to heart, some philosophers, Caputo for instance, have called for a 'new' Enlightenment:

> [T]he old Enlightenment has done all the good it is going to do and we now need a new one, not an anti-Enlightenment but a new Enlightenment. We have to...continue the Enlightenment by other means—to be enlightened about Enlightenment—to appreciate how much more non-programmable and inexact things really are.

> The idea is not to put out the light of the Enlightenment, but to put out a new revised edition by complicating its Pure Light with shadows, shades, greys, black holes, and other unexpected nuances and complications. (2013, 10)

This is vague as it stands, but I think something useful can be made of it. The two broad 'nuances and complications' in which I am primarily interested have to do with learning to appreciate non-linear complexity and extending moral cosmopolitanism so as to embrace future generations. Ultimately, I will define the Anthropocene Project in a way that brings these two considerations together. But since any apology for the Enlightenment, even one attentive to shadows, steps into something of a cultural minefield, I need to begin by being precise about which aspects of the Enlightenment I am *not* defending. Three are noteworthy.

First, I am not claiming that we should scour the *Encyclopédie*, for instance, in a vain hunt for specific solutions to our ecological crisis. What follows in this chapter is not a careful historical reconstruction of key Enlightenment texts designed to show that Diderot, Voltaire, Condorcet et al., were environmentalists *avant la lettre*. I am less concerned with what went on in the eighteenth-century world of ideas than I am with how we might take up those ideas for our own purposes. Nor, second, am I claiming that we should take on board every principle espoused by Enlightenment thinkers. Many of these figures were *principled* racists and sexists.[3] Third, and most important, I think we need to reassess without wholly rejecting the Enlightenment claim that progress in knowledge, both of nature and of society, is inevitable as well as the assumed corollary that such increases lead invariably to human betterment. Is anything left of 'the Enlightenment' after this winnowing? I think so.

To begin, we should embrace the Enlightenment's commitment to moral progress. The reassessment we require on this point has three hinges. First, we cannot be nearly *as* confident as Enlightenment thinkers were about the future steadiness of knowledge accumulation. Knowledge is not inevitably cumulative because its production is inextricably caught up in complex social structures and forces whose operations can impede the flow of ideas as much as advance it. Second, Enlightenment thinkers believed that knowledge is useful because it allows us to control its objects. But it can be just as important to realize that our knowledge and therefore our ability to control things in some domains might be limited in principle. Discovery of limits and uncertainties should, however, be counted a genuine epistemic achievement, an advance of enlightenment in some sphere. Finally, we need to be more precise, and less sanguine, than Enlightenment thinkers were about the connection between knowledge accumulation and human betterment. The latter happens, and we progress morally, only when explicitly moral considerations are treated as superordinate to developments in other domains, especially science, technology, and economics. I will develop these ideas in subsequent sections of this chapter, but for now we need to look more carefully at the concept of moral progress itself as well as two specific threats—the tragic outlook and pessimism—to our confidence in it.

What is moral progress? It occurs when some state of affairs is morally better than what preceded it. Following a number of writers on this topic, we can characterize 'morally better' in this formulation as increased awareness of and attention to the vital interests of genuine moral subjects or patients, those to whom moral agents might have duties (Jamieson 2002; Singer 1981; Rorty 2006; Williston 2011). We progress when we discover that groups previously thought to be outside the circle of moral concern—barbarians, slaves, women, some non-human animals, and so on—belong by right inside it. Let's call this the expansionist ideal of moral progress. Dale Jamieson has provided an 'index' of moral progress, a list of values that anyone committed to the soundness of the expansionist ideal should be able to endorse: for example, abolishing war and slavery, diminishing poverty, and improving the social lot of marginalized groups (2002, 12).

The question we need to ask is whether or not *we* should adopt a positive attitude to the possibility of moral progress, thus construed. If so, it must be reasonable to judge that, with respect to any of the items on Jamieson's index, some sort or degree of confidence that things will improve is warranted. But rather than asking about all of Jamieson's items, I want to focus on one: improving the lot of marginalized groups. And I will narrow the focus even further to examine whether or not we can justify a positive attitude about the chances of defending robustly the vital interests of future generations in our deliberations about how to organize our societies and economies in the age of climate change. Finally, because our carbon emissions are such a potent threat to the future, such reorganization must begin with the swift decarbonization of the global economy, so that is the practical task I will focus on (more on the nature of this task just below). Can we be confident that we will achieve this, or not?

As I will show in Chapter 3, we are not currently moving to any significant degree in this direction, so there is obviously much room for improvement. But as regards what sort of attitude we should adopt to this future there are two specific challenges to be met. The first questions the very possibility of moral improvement in any area, while the second allows for this possibility but denies that we are in a position to achieve it with respect to the climate crisis. The first challenge—the tragic outlook—arises from the alleged incomparability of our values. We cannot progress or regress morally unless the values we instantiate, or fail to instantiate, in our practices are diachronically comparable. But, so goes the challenge, they are *not*, and the tragic outlook is therefore appropriate for us. George Harris has recently elaborated a version of this claim, arguing that the tragic outlook is best equipped to deal with the problem of deep loss without the consolation of 'pernicious fantasies' (2006, 19). Harris's account is useful for our purposes because he insists on placing us in an historical narrative and then making judgements about whether or not we can make progress on the basis of the opportunities and limitations this placement presents us with. I think the construction of a moral narrative appropriate for the new geological age is key to our self-understanding, so this is a promising theoretical framework.

Where do we moderns sit in Harris's narrative? Two features of our place stand out. The first is that we are heirs to the horrors of the twentieth century. This is significant because it rules out nihilism. The nihilist denies that there is anything of deep significance in life, hence no particular reason to mourn the moral losses of the last century, nor indeed to view them as losses at all. The nihilist is objectionably superficial because she cannot recognize deep loss. But Auschwitz is a *reductio* of nihilism. The second key feature of our historical place is our value pluralism, the product of modern liberal-democratic political culture. For Harris this pluralism is deep and pervasive. However, taken alone it does not constitute a threat to the expansionist ideal. The real threat to this ideal comes from 'tragic pluralism', the claim that our values are incomparable.[4] This would mean that we cannot achieve what the expansionist ideal points to: a state of affairs in the future that is better than the one that obtains right now. The ideal depends on the ability to compare the present with the future and both with the past according to some covering value, and Harris does not believe such comparisons make sense.[5] Reason should therefore 'grieve' at its inability to light the way forward for us.

The example Harris invokes in support of this claim is choosing between saving a loved one versus saving a group of decent strangers (you can't do both). We assume (a) that the loved one wins for a small group; but (b) that at some point, the group gets large enough to tip the balance the other way. The problem is that there is no formula to determine where the line is here. This means that the two values—the life of a single loved one and the lives of an indeterminate number of decent strangers—are incomparable. And, importantly, this problem generalizes to all of our key moral choices in public life: balancing equality and excellence or liberty and security, for example (Harris 2006, 249). But is it correct to describe as incomparable the values at play in cases like this? In spite of a lengthy analysis of the issues just mentioned, Harris does not show that they are best viewed as clashes between incomparable values whose 'resolutions' are not intelligible. Instead, they appear to be difficult-to-resolve conflicts whose resolutions admit of reasonable, hence intelligible, compromise (by 'satisficing', for instance[6]).

But suppose I'm wrong about this. What follows if our values *are* incomparable to the degree Harris supposes? According to Harris there are multiple possible 'attitudes' for reason to adopt in the face of our moral conflicts:

> Reason's regret is the rational/emotional response to tragic loss where what is lost in the lesser good is not contained in the greater good. Reason's grief is the response to unintelligible loss, where what might be gained is incomparable to what might be lost. And reason's despair is the response to tragic loss when it is rational to think that the bad outweighs the good. (2006, 262)

Grief, the appropriate response (for Harris), thus sits midway between regret and despair. When our values clash, there are winners and losers. We do not have the consolation of understanding our losses (the standpoint of regret) but neither are we pushed to the view that the bad will prevail over the good (the standpoint of despair). The problem with this is that it is hard to see what Harris thinks could support any talk about 'moral loss' at all given his commitment to radical value incomparability.

Here, a distinction between moral loss and emotional pain might be useful. A person who cannot make sense of a conflict she encounters may experience moral loss if she has come to expect better of the world, if similar conflicts in the past had been resolved in a manner that was intelligible to her. In this case, she is, let's suppose, both pained by the event and able to experience it as a moral loss, a fall from expected patterns. But if our values are incomparable we will lurch from one choice-context to another. The lurching can be either pleasant or painful, depending on what is at stake. But there can be no experience of moral loss (or gain) because there is no rational expectation that things might have gone differently. The specifically moral distinction between is and ought collapses.

Our values are either utterly incomparable, in which case it makes no sense to speak of gains and losses over time with respect to those values, or they are not, in which case we can speak this way. But if we cannot, then the appropriate attitude to adopt is neither despair nor regret (both of which entail the possibility of comparison) but nihilism. In this case it looks like we are condemned to moral superficiality even though we might experience a great deal of emotional pain at what transpires in our lives. Harris is right to reject this view because it is an unacceptable response to deep moral loss (to any moral loss, really) but he fails to see that to abandon all value comparability is also to give up on moral seriousness.[7] Harris is therefore either a proponent of some value comparability and consequently a pessimist or optimist about the future, or he is a proponent of radical value incomparability and a nihilist *malgré lui*.

The second challenge to the achievability of moral progress is pessimism, the claim, as Harris puts it, that in the future the bad will outweigh the good. Again, although it is possible to be a pessimist about any or all of the items on

Jamieson's list, I want to focus on pessimism about whether we will avoid the worst effects of climate change by moving now to decarbonize the global economy. Here I want to defend the idea that we should avoid *acting* pessimistically. Let me distinguish here between epistemic and practical pessimism. Epistemic pessimism is the judgement that the bad will probably outweigh the good in the future with respect to some value or set of values. It is simply a probability-based belief. Practical pessimism is the judgement that, given epistemic pessimism about a particular outcome, we should *act* as though this is the case for that outcome. Crucially, epistemic pessimism does not entail practical pessimism in every case. There are two reasons for this. First, acting as though things will turn out badly might lead to a worsening of outcomes, and we have independent reason to avoid this. Second, an alternative attitude—rational hope—is compatible with epistemic pessimism. The full argument for the second claim must await Chapter 6, but I will spend most of the remainder of this section elaborating the first claim.

Before getting to that, however, more must be said about 'rapid decarbonization'. As I have said, this would be the single best practical expression of a commitment to de-marginalize future people in our deliberations about how to structure our society and economy in the age of climate change. But the term 'rapid decarbonization of the global economy' is shorthand for a very complex and multifaceted task. At the moment, the best tool we have for understanding this goal is the recently developed carbon budget. If we are going to avoid the worst outcomes of climate change, we cannot burn all the fuel we look set to burn. If we want a 50 per cent chance of not exceeding the '2°C guardrail' endorsed in principle at Copenhagen in 2009, our total carbon budget is 3700 billion tonnes of CO_2 (Berners-Lee and Clark 2014, 25). Unfortunately we have already burned half of that since the Industrial Revolution. So we are left with about 1600 billion tonnes of CO_2 on the 'coin-flip' scenario. A 75 per cent chance of avoiding 2°C cuts that amount (roughly) in half again, to about 700 billion tonnes of CO_2 (Berners-Lee and Clark 2014, 26).[8] The problem is that the proven coal, gas, and oil reserves of the world's fossil-fuel companies and petro-states amounts to 2796 billion tonnes (Berners-Lee and Clark 2014, xiv). The main task is therefore to constrain the supply of fossil fuels through the imposition of a global cap-and-trade system or carbon tax. This is by no means the only thing to do nor is it meant to foreclose all disagreement about the specifics of public policy.

A key consideration here has to do with the energy demands of the developing world. As I show in Chapter 3, by 2035 90 per cent of net growth in demand for energy will come from non-OECD countries. We likely cannot slow this demand—and even if we could it is probably not just to try—but it may not be necessary that the bulk of it be supplied by fossil fuels (though this is the current trend). Even if this is necessary, then the right of the developing world to a decent standard of living means that these countries should be able to boost their use of fossil fuels even as rich countries reduce theirs dramatically. The point is to bring net emissions down in accordance with what the science tells us. It does not follow that poor countries will be able to develop as profligately as we did, however. Henry Shue has shown that the grave realities of climate change entail that even the very poor will be asked to 'conspire in the imposition of limits on their own children's dreams' (2014, 71). This is of course compatible with seeing to it that everyone will receive enough for a decent life. Exactly how long developing countries will require fossil fuels to meet their basic needs is a contingent question, but there is a sense in which from the standpoint of the need for constraint the precise answer does not matter. That the total supply of fossil fuels needs to be massively constrained is an ethically inescapable fact.

It should go without saying that alongside aggressive constraint of the fossil-fuel supply, substantial effort must be put into research and development of renewables, and efforts to bring these technologies en masse to non-OECD countries must be significantly enhanced. Additionally, we should seek to reduce soot from cooking fires and methane from livestock and deforestation, at the same time as we develop carbon capture and storage technology (Berners-Lee and Clark 2014, 184–8). We've also got to think much more seriously about adaptation measures, especially in the developing world. Finally, when I say we must do these things 'rapidly' I am speaking in terms of targets aimed at defining an emissions peak followed by a reduction rate. The longer we take to peak, the faster we must reduce. So, for example, if we peak in 2015, we must reduce at 3.6 per cent per year thereafter, but if we wait until 2020 we must reduce at a whopping 6 per cent per year (and 12 per cent for a peak of 2025) (Schellnhuber 2009).

This, then, is the thumbnail version of what I mean by 'rapid decarbonization'. When we bracket all the social and political forces working against serious efforts to achieve it, we see that there *are* possible paths forward. I remarked that there is ample room for debate about the policy specifics. Consider, by way of example, the following set of policies designed to achieve rapid decarbonization. The International Energy Agency (IEA) (2013) has recently argued that we can limit twenty-first-century warming to 2°C above the pre-industrial baseline by adopting four strategies: large-scale targeted energy efficiency measures, limiting construction of coal-fired power plants, reducing methane emissions from upstream oil and gas facilities, and moving quickly to end subsidies to the fossil-fuel industry. My point is not to endorse this particular cluster of policy prescriptions (and we should be especially wary about efficiency measures, as I argue in Chapter 4), only to suggest that we would not bother to generate such strategies at all if we thought there were no viable paths to rapid decarbonization.

However, one might object that the case for practical pessimism rests on precisely the forces I have just bracketed. It is because fossil-fuel consumption is so entrenched in our societies and so bound up with the current structure of power that we will not pursue the necessary measures. The problem with this response is the feedback loop between our perception of a bad state of affairs and the ways in which the latter can worsen objectively as a result. Mediating between the first and second elements is what Christopher Lasch (1985) describes as a kind of 'psychic retreat', consequent on the pessimistic belief that things are bad and likely to get worse. As Joe Bailey has argued, this amounts to a refusal 'to imagine the future' (1988, 70). As such its most potent effect is to reinforce the current power structure, which works to close off futures contrary to its vision and interests.

There are two ways to elaborate this point. The first is that practical pessimism *colludes* in the production of negative outcomes. Of course, those who benefit from current power structures don't describe the outcomes this way. Many of them are Innerarity's 'accelerators', those who think the problems of the future can be handled through technological innovation and economic expansion. Possible social, cultural, and moral changes are either left out of the picture altogether or, more typically, are seen as epiphenomenal effects of the primary drivers: 'the possible future is by this means colonized by the local present and the crises . . . are left unconnected to both current political positions and possible cultural alternatives' (Bailey 1988, 89). That is, precisely to the extent that we adopt the practically pessimistic stance the crises we face are not seen as politically constructed and therefore contingent. They are instead cast as necessities. This both serves vested interests—why oppose what is believed to be necessary?—and is likely to hasten the arrival of the scenarios we fear. But an attitude whose widespread adoption has these effects *intensifies* bad outcomes by decreasing or removing constraints on the forces producing them. Since a bad outcome is rationally undesirable, the less of it we have the better. So we should reject whatever attitudes enable its increase.

Second, there is something objectionably self-indulgent about practical pessimism, at least as applied to the climate crisis. In her study of the concept of evil in modern philosophy, Susan Neiman argues that the moral task of the modern age is to refuse to 'feel at home' in the sociopolitical structures we have built. Commenting on Adorno's reflections on the role of practical reason after the Holocaust, she argues that decency demands 'we refuse to feel at home in any particular structure the world provides to domesticate us' (2002, 305). Here what I said about nihilism—that it cannot make a meaningful distinction between what is and what ought to be—applies equally to practical pessimism because of the lassitude it induces in us. The only difference between the two attitudes is that the practical pessimist may preserve a vestige of the distinction through fantasy, nostalgia, or cynicism. But he cannot honour it where it belongs: in the domain of practical reason. In spite of his possibly sophisticated and seemingly detached world-weariness he is entirely too comfortable in the house built and maintained by those in positions of social and economic power.

It might help to imagine how a person of the future would look at an early twenty-first-century climate change pessimist. If the effects of climate change have ravaged the future person's world, she is likely to see his practical pessimism as a response to a difficult challenge that is at best lazy and at worst cowardly. This judgement would gain significant force insofar as this person compared her society's choices to ours. For example, Catriona McKinnon has described a future world—the 5°C world I will look at in detail in Chapters 3 and 6—in which official decisions about life and death are a matter of 'triage'. That is, choices regarding who should receive resources necessary for survival are based on calculations about how long people are likely to live and what quality of life they will likely enjoy. Where there are competing claims among those equally efficient at 'converting resources into life', decisions will be made on the basis of a survival lottery.[9] Since they would likely see themselves as merely coping with disaster, it would be hard to criticize administrators who make these decisions—not to mention citizens who have to live with them—for being pessimistic about the future of their societies. Things do not have to be this bad to justify practical pessimism, but the comparison to our situation should make us question whether this stance is warranted for us.

2.3 Between Engulfment and Omnipotence

So we should indulge neither the tragic outlook nor pessimism, but are there other reasons to reject the Anthropocene Project? Jonathan Glover (2000) has recently examined the extent to which utopian fantasies played a key role in the atrocities of Stalin, Mao, and Pol Pot (among others). In calling for another 'grand narrative'—and it is hard to deny that the Anthropocene Project fits this description—are we not in danger of endorsing the same sort of morally repugnant teleology that drove these reigns of terror? Grand narratives are not, however, inevitably totalitarian or homogenizing. The civil rights movement was a grand narrative; so was abolitionism, and so is the ongoing battle for gender equality. These movements are, or were, grand in the sense that they put up profound challenges to existing power relations, possess international reach, and challenge the way we organize both public and private space. There is nothing totalitarian about these movements in spite of their great success in fundamentally altering human relations. Indeed, since their purpose was and is to expand the circle of moral considerability—so that new voices enter the moral and political dialogue—they are intrinsically pluralistic grand narratives. In his *Anthropology*, Kant defined pluralism as 'the attitude of not being occupied with oneself as the whole world, but regarding and conducting oneself as a citizen of the world' (Quoted in Lloyd 2013, 154). This makes it look as though we can have grand narratives based on cosmopolitan openness and pluralism.

The Anthropocene Project, as I conceive it, takes this possibility to heart. Totalitarian or homogenizing grand narratives are characterized by a pathological belief in our ability to control the social and natural worlds. Pluralistic grand narratives avoid this pathology through a willingness to live with Otherness, a refusal to bend what is strange and unpredictable into the straight lines of a single possible future while taking due moral responsibility for the effects of our actions. This new responsible pluralism applies in the first place to how we approach other humans, and in the Anthropocene this will involve, centrally, efforts to bring the claims of future people more actively into our public deliberative processes (via proxies). Since we now have a profound influence over the lives of these people we are forced to think hard about what we owe them. We are accustomed to investing for near and dear and are happy to allow our traditions and practices to pass smoothly down the generations. Now, however, it looks as though we must make deep and ongoing sacrifices on behalf of future strangers. Given that this will be a permanent alteration in our ethical framework—this is what makes it different from the same sort of commitment applied in wartime, for example—it entails a novel approach to policy formation. In later sections of this chapter I will explain the philosophical foundations of our responsibility for future people. But to set this up, we need to see how to reconceptualize the way we both shape and are shaped by the various social and natural systems in which our lives are embedded.

One of the key features of our new insight into systems is that we lack absolute control of them. Our social systems, precisely to the extent that they are infused with our technologies, are largely unpredictable. They are 'techno-social systems' (Allenby and Sarewitz 2011, 32). As Allenby and Sarewitz argue, we need to eschew simple cause and effect explanations of our purposes and the machines we invariably use to enact them. Take air travel as an example. Allenby and Sarewitz distinguish three interconnected levels of analysis of this seemingly simple phenomenon (and then generalize to all techno-social phenomena). At the first level, the jet carries one from points A to B with 'incredible reliability'. Mechanical explanation is adequate here. Level two describes the larger techno-social system in which level-one machines are embedded: in our example, the air traffic system, which is often extremely unreliable. This level is much less bounded than the first one: 'it includes subsystems—airline corporations, the government security apparatus as applied to air travel, and market capitalism in route pricing, to name a few—that, acting together, create emergent behaviours that cannot be predicted from the behavior of level [one] aircraft units' (Allenby and Sarewitz 2011, 37–8). Finally, level-three analysis includes the larger matrix of social, environmental, and personal effects that co-evolve with the phenomenon in question. With air travel we get 'significant changes in environmental and resource systems...mass-market consumer capitalism...individual credit...behavioral and aesthetic subcultures and stereotypes' (Allenby and Sarewitz 2011, 39).

Although I don't agree with the precise way in which the boundaries are drawn here, we can employ something like this model to conceptualize the manner in which our techno-social systems affect and are affected by climate change.[10] Here, level-one analysis might describe the complex, though still more or less mechanical, nature of the global carbon cycle and how anthropogenic interference in it works. Because we are describing a stable cycle, things are relatively bounded at this level. Although human interventions in this cycle are evident going back at least 8000 years, until recently they were not sufficient to disturb it radically.[11] Call this *linear intervention*. Level-two analysis would be required to explain how it is that the climate is prone to surprises in the form of positive feedbacks, emergent properties, and tipping points. There are purely natural reasons for such surprises to happen, such as runaway greenhouse conditions leading to mass extinction events, something which has happened at least five times in the history of life on earth.[12] But human systems are now the primary drivers of these phenomena in the climate system. This could not have happened absent a huge population applying its technology in a concerted and sustained way. Call this *non-linear intervention*. Because we can imagine this intervention not penetrating into every aspect of social life, we require another level of analysis to describe what happens when our climate-altering technologies and the social systems supporting them come to define virtually every aspect of social organization. This is level-three analysis, which looks at the vagaries of global economic and geopolitical forces, including global and local wealth disparity as well as population growth. It will also include the many ways in which our individual and group identities are constructed through and by our high-carbon lifestyles. Call this *pervasive, non-linear intervention*.

Let me make three points about this classification. First, at all three levels we are to some extent focused on the behaviour of a purely geophysical phenomenon, the climate. But the causal impact of techno-social forces on the climate becomes more and more convoluted at each step. As Allenby and Sarewitz note, the phenomenon itself becomes less bounded as a result. This makes it increasingly difficult to predict how it will evolve. Second, the intensification of techno-social intervention in the climate that takes place from linear intervention to pervasive, non-linear intervention is accompanied by an intensification in the feedback loop whereby the climate affects social systems. As we will see in Chapter 3, for example, this process may ultimately find expression in the violent fragmentation of many human communities struggling to cope with climate disaster. This can all get maddeningly complex because the fragmentation is both the effect of some original environmental degradation and the cause of increasing environmental degradation. Third, although we can choose any of these three levels on which to focus (depending on our explanatory goals), when we talk about 'the phenomenon of climate change' we must stress the importance of pervasive, non-linear intervention. This is because in the Anthropocene this form will become more and more prevalent and our most salient ethical choices will have to do with how we manage our way through them. It is imperative to note, however, that this focus does not license an 'anything goes' relativism about what climate change means. As I argue in Chapter 5, it is fully compatible with the pride of place we rightly give to climate science in our efforts to understand climate change.

For Allenby and Sarewitz, the dream of establishing full control over interdependent natural and social systems like these is sheer hubris. They conclude that such complexity is the 'assassin of Enlightenment ambition' (2011, 160). The analysis of complexity offered by Allenby and Sarewitz is impressive, but the criticism depends on the mistaken idea that level-one analysis is essential to the 'enlightened' view of knowledge. This is false. Enlightenment philosophers thought that enlightenment was a process, one capable of historical evolution. There is very little historical evidence for the claim that these thinkers believed in a fixed form of scientific explanation. In his still unparalleled study of the philosophy of the Enlightenment, Cassirer remarks:

> The thought of the Enlightenment again and again breaks through the rigid barriers of system and tries, especially among its greatest and most original minds, to escape...strict systematic discipline. The true nature of Enlightenment thinking cannot be seen in its clearest and purest form where it is formulated into particular doctrines, axioms and theorems; but rather where it is in process, where it is doubting and seeking, tearing down and building up... [I]t consists less in certain individual doctrines than in the form and manner of intellectual activity in general. (1951, ix)

In this spirit, the mark of Anthropocene intelligence is to grasp that modern technological interventions will inevitably reverberate in directions and magnitudes we could not have foreseen. If in the face of this onslaught simple mechanical explanation loses its force in some domain, then science will craft new modes of explanation for that domain, and this should be seen as an evolution of scientific enlightenment. This is why Genevieve Lloyd has argued that the editors of the *Encyclopédie*—Diderot and D'Alembert—conceived of their project in an open-ended way. They took themselves to be constructing principles of knowledge that would contribute to a 'circle of learning' comprising past, present, and future truth seekers (Lloyd 2013, 125). What was crucial to the project was not this or that mode of explanation, but the idea that scientific knowledge should be allowed to accumulate without interference from religious authorities. There is no reason to doubt that these philosophers would have seen advances in ecology and systems theory—including the new appreciation of complexity and uncertainty—precisely as achievements of enlightenment principles. Again, Cassirer: 'only in *process* can the pulsation of the inner intellectual life of the Enlightenment be felt' (1951, ix; my emphasis).

The cardinal rule of the Anthropocene is to appreciate that we can no longer neatly separate the workings of techno-social and biogeophysical systems. Our most general task is therefore to manage their complex interactions in ways that are ethically justifiable. Part of what it means to be enlightened in the Anthropocene is that one approaches complexity, thus understood, neither fatalistically nor with fantasies of full control. To borrow terms from Anthony Giddens' analysis of modernity, this represents a refusal to be dominated by feelings of either 'engulfment' or 'omnipotence' in our efforts to put a mark on the world, and to try instead to stake out some middle ground between these two options (1991, 193–4). In the Anthropocene we will have to learn that our highest task is not to try and implement a fully preconceived plan but to manage risk wisely. Innerarity claims that this is what makes politics so important now. But enemies of the ideal exist on both the left and the right of the current political spectrum. The problem with the left is that it is mired in anti-realist utopian thinking (the preconceived plan), while the right is entirely uncritical in the face of the technological and economic juggernaut.[13] Our politics is predictably limp as a result: 'reality and efficiency are managed by the right wing, while the left is free to enjoy the monopoly of unreality . . . In this way, there are those who are afforded reality without hope and others, hope without reality' (Innerarity 2012, 121–2).

But there must be a limit to the rule of chance, and this is the part of the Anthropocene Project that emphasizes our ability to shape the future. Neither Innerarity nor Allenby and Sarewitz—nor indeed Ulrich Beck (2007) who got everyone talking about the risk society in the first place—is clear about what it means to 'manage risk wisely'. They do not see that this is primarily a task for philosophical ethics (an oversight for which they are blameless since they have received little help from philosophers addressing the problem). For his part, Innerarity seems to believe that the free play of an ideologically uncontaminated politics will all by itself allow morally decent solutions to our problems to emerge. But ideology abhors a political vacuum. It will not stay away simply because we wish it to. What exactly is going to prevent one or another sectional interest from colonizing the future to the possible detriment of most people of the future? What we require is a specifically moral constraint on political decision-making and policy formation.

How does this relate to what we have been saying about moral progress? One way to make the connection is through a more focused critique of the accelerative mindset. This makes sense because this mindset belongs squarely to the global elite currently perpetuating the fossil fuel energy regime. Those who claim that all our problems will be solved by technology and the market want us to simply leave these forces alone. The less constraint they have from above, the more likely it is that they will deliver the goods. The point of calling the techno-economic imperative a 'juggernaut' lies in the fact that this is the way many people, including its boosters, actually experience it: as a huge and unstoppable force to whose dictates we must blindly sacrifice ourselves. The best way to think about this is in terms of the concept of alienation. Although Marx elaborates the concept in a manner specific to capitalist modernity, its roots go back Plato's *Republic*, in the description of the ideal fit between the soul and the city (Marx 1964; Plato 1992).

At its most general, the push to de-alienate social life is aimed at structuring our institutions so that they reflect us, so that we can see our work, our values, and our aspirations in them. An unalienated social life is one in which we experience ourselves as free shapers of the conditions of our existence. Alienated social life, by contrast, is that in which such reflection is lacking. The social world appears not only alien in the sense that something other than our free choices appears to be behind it, but, because of this, it takes on the character of necessity. We feel helpless in the face of its demands. Investigating alienation is philosophically important because, as Marx and Engels saw so well, it is a principal source of the malaise of the modern. Constraint as such is not an evil, but to feel constrained by forces we neither understand nor control, especially where we think we *ought* to be able to exercise some control, thwarts our sense of ourselves as free beings. As Vogel has argued, alienation in this broad sense might be the key feature of the climate crisis. We are not, he claims, alienated *from* nature (as many contend), but from our already-actualized transformative presence *in* nature (2012, 306–7). The main reason for this for Vogel is that the capitalist system privatizes our choices, thus cutting us off from wider communities (2012, 309–10). This phenomenon is a key causal agent in the production of collective action problems. The task therefore is not to cease transforming nature—that is not even possible at this stage of the game—but to ensure that our transformations are the free and conscious product of a community of equals.

The difficulty at which we have arrived can therefore be expressed in two key points. First, our ideal of moral progress—and of the pluralistic grand narrative that expresses it—must be specifically designed to counter the alienating effects of the techno-economic juggernaut. Because our world is, as

Giddens claims, 'increasingly ordered according to the internally reflexive systems of modernity' (1991, 165), we need to construe our efforts to progress morally as superordinate to developments in economic and technological systems, and this entails that the latter must become objects of our conscious control to an extent we have not yet achieved. As it is, standards of progress in these domains are determined by those domains themselves. But because they have largely escaped our control, what we desperately need is a way of subordinating them to an external standpoint that is, as such, capable of shaping the way they develop.[14]

A pertinent example of the need for such control is our fascination with geoengineering . As Gardiner (2010) has shown, to the extent that we have accepted a choice between catastrophic warming and recourse to dangerous geoengineering methods we have already framed the issue in a question-begging way. This is because one of the two choices available to us assumes that mitigation won't work and that climate catastrophe is therefore coming willy-nilly. With this assumption in place, characterizing geoengineering as 'the lesser evil' effectively forces us to choose it. But why do we insist on this framing rather than working tirelessly at mitigation so that catastrophe does not become one of two evil options? In place of an honest assessment of the deep and manifold harms geoengineering is likely to bring to future people, we have been guided in our understanding of the issue by faith in technology's saving power.[15] This faith has been supported by moral weakness cleverly disguised as realism. We may ultimately need to resort to geoengineering but we should not do so before the full ethical implications of these technologies have been assessed, something that has not yet been done.[16]

But, second, the ideal of moral progress must be instantiated in a way that is optimally flexible in the face of the Anthropocene's uncertain and open character. There's no point in placing arbitrary or ideologically based constraints on forces whose peculiar energies and intelligence we require if doing so hobbles us in a world full of surprises. And there's a good deal of truth in the claim that those energies and that intelligence are more likely to emerge in and through the systems' own principles of organization. *If* it is the case, for example, that capitalism, suitably reformed, is the best economic system we have for bringing about the transformation to a decarbonized global energy regime, then we had better give it *some* freedom to do so. This should not be seen to invite complacency about contemporary capitalism. We can reject its growth imperative, its plutocratic tendencies, and its easy externalization of social and environmental costs. But we should also avoid heavy-handed constraint of this system if in some form it is the best one we have.

For example, there is some reason to believe that a global scheme for limiting carbon emissions through cap-and-trade might help push us quickly away from reliance on fossil fuels if it is constructed on the 'contraction and convergence' model (Brown 2013, 163–5). Although this scheme involves the top-down assignment of carbon permits—in line with what science tells us we need to do to meet certain temperature targets, for instance—the market takes over after that in determining how permits are allocated. Ensuing high prices for the permits could be a cause of strife to poorer countries and consumers generally, but this would not necessarily happen; and even if prices did rise, it could hasten the market-driven transition to renewables (Berners-Lee and Clark 2014 140–1).[17] This is far from establishing the case for cap-and-trade, of course, but on the assumption that some such scheme could achieve decarbonization more quickly than any feasible alternative, it would be foolish to oppose it on the grounds that it is not sufficiently anti-capitalist.

These examples—geoengineering and cap-and-trade—help define the paradoxical task of an unalienated Anthropocene: to increase significantly our control of what are now purely internally reflexive systems without undermining the ability of those systems to be flexible and creative in responding to unpredictability. Again, this 'control' should take the form of cosmopolitan ethical constraint, which must define the boundaries of permissible interventions in the systems that frame our lives. Anything else would allow for the colonization of the future by yet another sectional interest. Should we be optimistic about our chances of rising to the challenge of the Anthropocene thus construed? It is too early to say. With the tragic outlook and pessimism off the table, *some* more positive attitude toward our future is warranted, even while staring the climate crisis squarely in the face. But this makes available more than one possible attitude. Optimism is one, hope another. In Chapter 6 I argue that a certain kind of hope is justified here, but there's much work to be done before we are in a position to appreciate this claim.

2.4 Moral Cosmopolitanism

The more immediate task is to look carefully at the nature of the 'ethical control' to which I have been referring. In negotiating a path between the fantasy of omnipotence and subjection to fate as a specifically moral problem, it can be helpful to think about the role we assign to luck in our moral practices as well as our moral theories. This is a recurrent theme in contemporary moral philosophy, in part because it responds to a deep concern that we not exclude individuals or groups from the circle of moral considerability for morally arbitrary reasons. Since nobody is responsible for his gender, skin colour, place of birth, species membership (etc.)—these are all the product of natural and social lotteries—nobody should be disadvantaged in the distribution of burdens and benefits because they possess one of them or fail to possess another. When we seek to expand the circle of moral considerability in this fashion, we are in effect *taking control* of the institutions we use to distribute benefits and burdens in society through prescriptive specification of their normative foundations.

For all their differences about exactly how benefits and burdens ought to be distributed, diminishing the rule of the arbitrary is something all plausible normative theories accept the importance of. The differences do matter a great deal of course, precisely because they can sometimes reflect deep disagreement about what features are and are not matters of luck. A utilitarian and a libertarian might, for instance, disagree about the extent to which a person's being a member of an economically marginalized group is arbitrary. This matters in deciding whether or not the state is morally required to use some of its resources to help alleviate this person's condition. But if two theories—or for that matter two people—agree that a person's possession of (or lack of) a given feature *is* purely a matter of luck, they will also likely agree that the person ought not to be systemically advantaged (or disadvantaged) solely on this account. This is virtually a test for plausibility in a moral theory or decency in a person (though Section 2.5 of this chapter takes up an important caveat). Still, we should recognize that at any point in our moral evolution we are unlikely to have a complete answer to the question about who or what has moral standing (Schönfeld 1992). So some entities that should be counted are not being counted, which means that they are being excluded from moral consideration for arbitrary reasons.

For example, when slavery was a common practice with little or no opposition to it, the idea that slaves might be morally considerable seemed unthinkable. They were chattel, pure and simple. It was not strictly necessary to add that their having been born to this fate was necessary. If pressed on the point a defender of the system might instead simply assert that although it is unlucky for anyone to be been born into slavery, no slave is as such treated unjustly, since our slave-holding system is itself morally unimpeachable. In other words, there is nothing *morally* arbitrary about being discriminated against in this way even if it is naturally arbitrary that one was born with dark skin, or in this recently conquered country, etc. Now, moral progress consists in large part in rejecting this train of thought, in seeing moral arbitrariness where we had previously seen only 'natural' variation. But why not count this as a discovery about ourselves?

When we make progress we should not say that we have simply changed our minds about what belongs inside the circle of moral standing but that what we were doing before was unjust by our own lights. And in the case of slavery this amounts to saying that, given our own principles, we should have recognized that skin colour is a morally arbitrary feature of humans. At this point, we will likely ask searching questions about what took us so long to make the connection. Now, do we know that we ourselves are not in a position similar to that of the defender of slavery? If we are, then we are excluding some entities from moral consideration on morally arbitrary grounds. This is a possibility to which we must always remain open in the unpredictable Anthropocene, as our technological reach is extended in strange new ways: to other planets, the longevity of our bodies, the farther future, the enhancement of robotic minds, etc.

One reason to think we might be in this position, for example, is that sentient non-humans, non-sentient non-human animals, plants, ecosystems, or the Earth itself (Gaia) may be morally considerable things. Indeed, with the exception of sentient non-humans, it seems to me that we may be in the dark for some time about all these entities. For instance, some people take being alive as such to be morally important. The various ways in which life can be expressed—the morphology of this or that living thing—is, these people think, morally arbitrary and cannot justify disadvantaging one kind of living thing relative to another. They might, further, have a very broad idea of what life is. So they might say that ecosystems are morally considerable. Since the way ecosystems cycle energy homeostatically makes them functionally like organisms it is and always was unjust to disadvantage them for the morally arbitrary reason that they lacked certain *other* qualities: reason, sentience, the easy individuation many living things enjoy. Given our full-out industrial assault on ecosystems, and the prima facie plausibility of the homeostasis-based analogy, this criticism surely has *some* bite. If the standpoint it expresses became the evaluative norm in some society (or in our own in the future), our critics might look with astonishment at our general unwillingness to put this—to them rather *obvious*—moral notion into political practice.

I raise this example because some might be tempted to say the same thing about our treatment of future generations. Here, the problem is in one sense easier to resolve, or at least approach, than it is with ecosystems, Gaia, etc., but in another sense it is just as difficult. It is easier because we are talking about humans but just as difficult, perhaps more so, because of their specific modal status. *How* to count merely possible humans in our moral deliberations is indeed a difficult question (which I partially address in Chapter 6), but it is important to begin by asking *why* we ought to count them. Following Onora O'Neill we might say that we should include in the circle of moral considerability any group whose vital interests are affected by what we do: 'those whom agents already take for granted in acting' (1996, 4). This sounds right because, having already decided that humans are equally morally considerable, the moral-epistemic task is to determine if we are excluding people our attitudes and actions suggest we should not be. If we were doing this we would face a problem of rational consistency: we would be supposing something at the level of our principles that we were in effect denying at the level of our attitudes and actions. To preserve integrity we would in this case be forced either to alter the relevant principles or change our attitudes and behaviours.

Moral cosmopolitanism is the view that every human is equally morally considerable and that we therefore have duties to other humans precisely to the extent that their interests are bound up with or taken for granted in our actions. This is the basis for the notion that we are, first and foremost, 'citizens of the world' rather than this or that political community. Before human activities on one side of the planet could affect people on the other side it did not make sense to say that the first group could have duties to the second. But things have changed with globalization and its effects. Now I can consume products made in faraway sweatshops or add to the already dangerously high atmospheric stock of carbon by driving to work. Whichever humans are affected negatively by these activities deserve more attention than people like me standardly give them.

Consider again Giddens' understanding of globalization in this context: he argues that globalization 'stretches' the local so that it encompasses the distant (1990, 64). If this is correct and we add to it O'Neill's understanding about how to fix the scope of moral consideration, it follows that globalization opens up new domains of ethical concern. This type of thought is part of the original impetus of Enlightenment cosmopolitanism. People in the early-modern era discovered that there was a world of humans beyond Europe which the latter's practices—war, exploration, colonization, and trade—were now impacting in morally considerable ways. And to some extent the thinkers of the day, if not always its people or princes, responded to these novel circumstances by arguing that state boundaries and the loyalties they impose were becoming less important.

But what we should say now is that the stretching process to which Giddens refers applies as much temporally as spatially. The agent of temporal stretching is a suite of gases—CO_2, CH_4, N_2O, PFCs, and HFCs—which spread uniformly across space and persist through generations. They are the ultimate boundary-busters. Because it has gotten so high by recent historical standards the atmospheric stock of these radiative forcers now threatens to disrupt the climate system dangerously. We are the major causal source of this phenomenon the effects of which will negatively affect the vital interests of people of the future. It follows that we can no longer exclude such people from the circle of moral considerability. Moral cosmopolitanism, and the conception of practical reason it embeds, *must* be extended to recognize the fact of this new temporal stretching. In what remains of this section I want to explore the sense in which we can see this challenge as one that is historically quite specific, that is, one that defines us as members of the Anthropocene epoch.

There is a particular philosophical worry behind this exercise. The worry, expressed by O'Neill, is that it is difficult to vindicate a conception of practical reason that is by its own lights historically particular. In my view, Korsgaard gets things exactly right in describing the issue this way:

> The conception of *moral* wrongness as we now understand it belongs to the world *we* live in, the one brought about by the Enlightenment, where one's identity is one's relation to humanity itself. Hume said at the height of the Enlightenment that to be virtuous is to think of yourself as a member of the 'party of humankind, against vice and disorder, its common enemy'. And that is now true. But we can coherently grant that it was not always so. (1996, 117)

Since this view presupposes a historically particular conception of practical reason, it would follow from O'Neill's claim that it probably could not be vindicated. Historicism (itself a form of particularism) 'locates the source of practical reasoning in the (supposedly) actual norms or commitments of a given time or place' (O'Neill 1996, 49). However, O'Neill doubts that this view can show that the ethical standards it offers are 'justified as reasonable'. This is reiterated in her claim that, although the task of philosophical ethics is to provide this-worldly foundations for practical reason, and thus to reject metaphysically or religiously based appeals, historicism—articulating 'historically achieved ethical standards and practices'—fails because it cannot 'build to suit our needs, rather than trapping us in the buildings and boundaries of a given time' (1996, 212).

There are two responses to make to this. The first is that O'Neill's Kantian account is itself more historically bounded than she evidently supposes. As we have seen, we cannot disavow principles that are assumed in our activities. If others are causally connected to us then they are by that fact moral subjects for us. Conversely, if there is no connection there is no need to include them. So, to take O'Neill's example, the T'ang Chinese and the inhabitants of Anglo-Saxon England, because they had no causal contact with one another, were not morally required to consider one another as moral subjects (1996, 101). O'Neill would surely agree that the economically and technologically determined geographical reach of these two groups were contingent facts about them, as were their respective social practices and standards of evaluation. If historicism is the view that 'historically achieved ethical standards and practices' are our moral guides, then in ignoring one another because of contingently evolved standards and practices, the T'ang Chinese and the inhabitants of Anglo-Saxon England were fully moral by historicist lights. However, the very same thing can be said of the Enlightenment ideals O'Neill endorses. These ethical standards and practices are as much a product of contingent historical circumstances as are those of previous groups of humans. At least, O'Neill has given us no reason to doubt this.

But, second, why should we consider ourselves 'trapped' by this? We can build only with the 'supplies we have been given'. True, and the Enlightenment has given us useful supplies, but we should not imagine that they could belong to any but the inhabitants of a particular historical time. Other, more particular, historical forms can trap us because they have only internal standards by which to vindicate themselves, and O'Neill clearly thinks this is inadequate. Can the same thing be said of cosmopolitanism itself? That is, can its principle of universal inclusivity be vindicated only internally? I see no reason to deny this, but O'Neill simply ignores the question, assuming perhaps that it does not arise for her account. It clearly *does*, however, even though the admission counts as a defeat only by comparison to a sky hook ideal of vindication, which we have no reason to accept. If ethics is intrinsically practical, the only question we need to ask is whether the tools we have to hand allow us to do the job that needs doing. And what needs doing, among other things, is attending to the vital interests of future generations. O'Neill notes that cosmopolitanism arose at a time of increasing integration, and the key question was how we could live peaceably and justly in a world of gradually dissolving or expanding borders. These changes provoked an intellectual crisis in early-modern Europe, the ethical response to which was a move away from traditionalism and towards cosmopolitan universalism. Our situation is similar and the tools used to address the crisis then are still, by and large, available to us.

2.5 Moral Luck and Generational Membership

Nevertheless, we might think that cosmopolitanism is too weak to draw future generations into our moral deliberations in a substantive way. Even though we recognize the moral considerability of future people, their interests might not override the special duties we have to members of our own generation. Consider, in this connection, the analysis some cosmopolitans make of our attachments to the nation (Tan 2004). It can be argued that being a citizen of this or that nation is morally arbitrary in the same way any other feature of an agent might be—her height or shoe size, etc.—and that such features cannot play a role in the specifically international assignment of benefits and burdens. It might be useful to see how some have defended national membership against this challenge, then go on to see if the same considerations apply to generational membership. If successful, the defence thus extended would show that relatively robust generational favouritism is warranted.

The first thing to note is the qualification we must make to the idea that any feature possessed only by luck is always morally arbitrary. Someone who has been handicapped from birth has different needs than someone who is not handicapped in this way. For David Miller, while it is true that neither the handicapped person nor the non-handicapped person is responsible for his condition, many would insist that they ought to be treated differently in the distribution of benefits and burdens. So, something that is true of an agent only by accident *can* give her a claim to greater resources than another. We would certainly not say the same thing about hair colour, which is also something for which we are not responsible. So is national membership like hair colour or differential need?[18] Miller argues that it is more like differential need, a claim that justifies privileging compatriots over strangers in some resource-distribution conflicts between the two groups.[19]

This argument rests mainly on the claim that so long as a national group in its 'present form' is not inherently unjust, favouring it in some conflicts is permissible (Miller 2010, 385). Miller agrees that insofar as the mode of existence of a particular in-group directly disadvantages members of out-groups, the latter need the protection of rights against the potential depredations of the former. But he does not think we need to go beyond this minimum, from which it follows that we can have special duties to compatriots. In particular, while we are not permitted to kill or injure non-nationals, we are not required to secure their basic rights when third parties have infringed them or when they themselves are responsible for their own deprivations. This means that in a world of scarce resources, we can, say, continue to fund quality education for our compatriots rather than using these resources to come to the aid of a foreign country whose citizens have been deprived of some basic goods by a third-party aggressor.

How does all of this apply to generational membership? Notice, first, that we need not determine whether or not generational favouritism as such is unjust but only whether it is so in its present form. I suggest that given the fact of climate change, the manner in which the present generation gets most of its energy is not an incidental fact about it but something essential to its present form. Second, we need to add to Miller's category of impermissible acts those that can injure other parties *indirectly*, in our case through the degradation of their physical environment.[20] But, third, notice that, in contrast to the conclusion Miller draws about the limits of our obligations to non-nationals, these points lead to a very *strong* statement of our duties to future generations. This is because the scarce resource at issue is the absorptive capacity of the atmosphere (as well as other carbon sinks, like the ocean). We are either close to or have already exceeded the point of carbon saturation, such that a further increase constitutes 'dangerous interference' in the climate system.[21] But if this is correct, and if continuing our way of life depends on adding even more carbon to the atmosphere (which it does), then in order to comply with our minimal duty of justice not to injure future people if we can avoid doing so, we must cut our use of fossil fuels *drastically*. This means, among other things, decarbonizing the global economy as quickly as possible.

These considerations reveal a key assumption underlying Miller's analysis, namely that while he certainly does not suppose resources are infinite they are assumed to be relatively abundant. This is why we can fund high-quality education in our country without killing or injuring non-nationals—to say nothing of nationals—to pay for it. But what if this assumption became untenable? Tim Mulgan (2011) argues that people in a 'broken world'—that is, a future profoundly stressed by climate-induced challenges—might have a difficult time understanding a good deal of our moral and political philosophy for just this reason. The severe scarcity of resources with which such people must cope figures prominently in both the design of their institutions (they have a survival lottery when times get really tough) and their moral and political theories (they are strict utilitarians). They might wonder why, given our clear intergenerational duties, we did not realize that we were in fact facing a scarcity of relevant resources more dire than we took it to be. Some of that absorptive capacity we were using belonged by right to them, but because we did not recognize this (nor take significant steps to develop alternative energy sources) they are now facing widespread and debilitating energy scarcity.

Being a member of this or that generation is entirely a matter of luck. It does not follow from this premise alone that we do not have special duties to members of our own generation, just as it does not follow from the fact that I did not choose my family that I have no special duties to my family members. But the duties I have to my family members, duties of loyalty for example, do not always override the duties I have to non-family members when the two conflict. If I'm born into a mafia family and an uncle 'asks' me to hush-up a murder, it should be pretty clear where my duty lies. The same goes for generational membership. When we humans did not have the capacity to injure people born a century or two after us our special duties to members of our own generation could go unchallenged. Widely separated generations were the temporal equivalent of the T'ang Chinese and the inhabitants of Anglo-Saxon England. Things are different now: because of how we get our energy we can't fund all the things that make our lives good without injuring future people. Not only is it the case that emissions are now zero-sum, but if we are to reduce them as rapidly as I have suggested we must, then we face what Henry Shue calls a 'shrinking zero-sum'.[22]

A distinction is sometimes made between moral cosmopolitanism and political cosmopolitanism (Pogge 1992). While the former is focused on fixing the scope of moral concern, the latter is concerned with specifying which institutions best express our self-understanding as citizens of the world. On this reckoning moral cosmopolitanism has 'no direct institutional implications' (Miller 2010, 379). It might be true that we can fulfill our intergenerational duties via more than one institutional arrangement (but maybe not). Even so, our analysis definitely has direct implications for the legitimacy of current institutions. To the extent that any of them—our economic system, the nation state, the United Nations, even the family—conspires, however tacitly, in breaking the future, it is unjust and must therefore be reformed or replaced.

2.6 Conclusion

I have described the Anthropocene Project as a culminating moment or phase of the Enlightenment project. It is important to do this because the idea of the Anthropocene can otherwise look like a radical interruption of our cultural history, making it difficult to know how to get from here to there. Michelle Moody-Adams argues that moral revolutions are often slow to come, not because we always lack the theoretical resources to understand how they might be justified but because of various cultural and socio-psychological obstacles preventing us from seeing our ideals clearly. These 'entrenched patterns of situational meaning' are a kind of thicket—composed of our habits (of consumption, of assigning specific emotional valences to happenings, etc.), the power relations that define our social and political lives, our virtues and vices, the metaphors we use to describe the world, and much more—that acts to obstruct our view of principles and ideals containing alternatives to our present reality (Moody-Adams 1999, 170).[23] In this chapter I have tried to peer *over* the thicket to the view beyond. My purpose is not to reveal something extra-cultural but rather another aspect of our culture, one that properly appreciated can help us discern which bits of the thicket to preserve and which to cut back.

This is to take seriously Kant's famous distinction between living in an enlightened age and living in the age of enlightenment (Kant 1988, 465–6). To the extent that enlightenment materials are available for our use we live in the age of enlightenment, though in failing to fit them fully to our purposes this is not an enlightened age. By way of summary, and also to set up the ensuing discussion, let's be more precise about the nature of these materials. The key figures of the Enlightenment were, I submit, motivated by three principal ideals or concerns. The first was justice, which for them was intrinsically bound up with the radical equality of all humans and the way this insight ought to structure our moral and political commitments. The second was the love of truth, a vigorous plea for scientific and artistic inquiry unmolested by power's capricious workings. And the third was hope, the confidence that reason can show us a way to improve social relations and thus allow us to live more fulfilling lives. These concerns were not merely theoretical. Though they are not always characterized as virtue ethicists, this is the most accurate way to describe what these thinkers were up to in their practical philosophies. They were clearly concerned with what it takes to produce societies of just, truthful, and hopeful *people*. This, I am claiming, is the essence of Enlightenment virtue ethics.[24]

We have not faced the climate crisis squarely in large part because we have failed in respect of these three virtues, and the only decent way forward for us is to learn, or relearn, how to be genuinely just, truthful, and hopeful people. In the next chapter I focus on a specific threat from climate change, one that has been relatively under-discussed in the climate change literature generally and ignored altogether by philosophers: the potential proliferation of serious inter-group violence as a response to worsening environmental and social conditions. The reason for this focus is not to downplay the other threats posed by climate change but to point to the sense in which anthropogenic climate change represents an indirect repudiation of the cosmopolitan ideal. Cosmopolitanism has often been seen by its supporters as entailing the rejection of warfare (except in genuine emergencies), because warfare, and violence generally, represent a disintegration of the human community (Nussbaum 2010). If the cosmopolitan ideal, suitably expanded in the way I have done in this chapter, is the moral heart of the Anthropocene, then climate-induced tribalism is perhaps the gravest threat to that project's integrative impetus.

3

The Spectre of Fragmentation

3.1 Introduction

In the St James's Palace Memorandum of 2009, a group of Nobel laureates invoked Martin Luther King Junior's phrase 'the fierce urgency of now' to describe the climate crisis (Quoted in Hamilton 2010, 196). They were worried that if we did not act decisively to mitigate our greenhouse gas (GHG) emissions, we were inviting a future of social and political chaos. In Chapter 1 I quoted IPCC Chairman Rajendra Pachauri's claim that 'the very social stability of human systems could be at stake' if we fail in this way, and it is now time to put flesh on this concern. In this chapter, I will therefore describe what the potentially violent fragmentation of the human community over the coming decades might look like. I will argue that the cosmopolitan ideal described in the previous chapter is being challenged by this potential state of affairs so that our inertia on climate change represents a repudiation of some of our deepest values.

I set out by describing what the latest science tells us about the climate path we are currently on, and look set to stay on, and argue that the evidence justifies our thinking explicitly in terms of looming catastrophe. This is articulated as a criticism of two mainstream social enterprises—economics and climate science itself—whose representatives have, in large measure (though in different ways), refused to speak this way. They have thus so far missed a chance to help legitimize a certain way of framing the issue for the larger culture. The physical science is one thing, its likely impacts on our societies another, and since one of my principal aims in this book is to lay out the latter threat as starkly as possible, I turn next to an examination of some recent work on the securitization of climate change. Here I focus on a particularly volatile region of the world, a huge swath wrapping the planet's middle and extending at least 35° north and south of the equator. The goal here is to show that on our current path, we are moving headlong to markedly increased inter-group violence in this area, possibly to a new age of tribalism. Indeed, connecting the securitization of climate change to new research projecting dates for future climate 'discontinuity', I argue that we are flirting dangerously with the possibility of sharp increases in crimes of atrocity in the decades and centuries to come. Finally, I explain why we need to both deflate the concept of security as applied to the climate crisis and insist on the importance of stable virtues in ourselves and others in facing the challenging times to come. This part of the chapter thus sets the stage for my analysis of particular virtues, which I undertake in the following three chapters.

3.2 Embedding Catastrophe

Like previous IPCC reports, AR5 projects various climate pathways we might take into the future. In previous reports, the pathways used qualitatively uniform socio-economic assumptions—about population growth, speed of transition to renewables, productivity and economic growth, and energy and materials demand—then tracked quantitatively distinct variations of these into the future, making the appropriate calculation about our GHG emissions, ensuing temperature rises, and effects on natural and human systems. For AR5, a modelling tool known as the Representative Concentration Pathway (RCP) replaces these scenarios. The main difference between RCPs and previous scenarios is that the former jettison the use of uniform socio-economic assumptions across the possibilities. Each RCP instead uses a set of socio-economic parameters specific to it. For our purposes what is important is determining the path we are on, and this looks to be at least close to the worst-case scenario as described by RCP 8.5. The latter is, simply, a 'non-climate policy scenario', which as such predicts a temperature rise anomaly in 2100 of 4.9°C relative to the pre-industrial baseline.[1] In this sense, RCP 8.5 is basically equivalent to A1F1, a previously used scenario involving rapid economic growth, a world population that peaks around 2050, and, most importantly, a fossil-intensive energy mix (IPCC 2007).

There are two reasons for adopting this scenario in diagnosing our current reality. First, as the 2009 *Copenhagen Diagnosis* makes clear, this is what the data about the key climatic indicators now show us. The authors of this report argue that '[a]t the high end of emissions, with business as usual for several decades to come, global mean warming is estimated to reach 4–7°C by 2100, locking in climate change at a scale that would profoundly and adversely affect all of human civilization and all of the world's major ecosystems' (Allison, Bindorf, et al. 2009, 49–50). They continue:

> Recent observations show that greenhouse gas emissions and many aspects of the climate are changing near the upper boundary of the IPCC range of projections.

Many key climate indicators are already moving beyond the patterns of natural variability within which contemporary society and economy have developed and thrived. These indicators include global mean surface temperature, sea-level rise, global ocean temperature, Arctic sea ice extent, ocean acidification, and extreme climatic events. With unabated emissions, many trends in climate will likely accelerate, leading to an increasing risk of abrupt or irreversible climatic shifts. (2009, 6)[2]

Projecting these trends forward accurately depends on assumptions about the policy choices we adopt in the near future, which brings us to the second point. The key to the A1F1/RCP8.5 scenarios is that they predict no substantive policy measures in the short to medium term moving us away from our reliance on fossil fuels. This looks accurate given our current commitments, at least on the all-important energy supply front. In its 2013 energy outlook the International Energy Agency (IEA) predicts that global energy demand will increase one-third by 2035. Even though the share of fossil fuels in the energy mix drops by 6 per cent in this period it still remains a very high 76 per cent. Obviously, the 6 per cent drop is not nearly enough to make up for the 30 per cent increase in demand. Moreover, by 2035, 90 per cent of net energy demand growth will come from non-OECD countries and in these countries the use of coal—the most carbon-intensive fossil fuel—will rise by one-third (though it will fall by one-quarter in OECD countries). Oil and gas companies—whether private or state-owned—are certainly *not* looking very far beyond 2035 in deciding which resources to exploit nor, evidently, are they particularly concerned about climate change. In fact, they are scrambling to meet the projected demand growth by locating and extracting any cache of oil or gas they can get their hands on, whether it is in the deep waters of the Arctic or off the coast of Brazil, beneath the thawing permafrost in Siberia, in Norway's far north, trapped in North Dakota shale rock, or in the Athabasca tar sands (Klare 2012).

It is illuminating to think about the trend shifting energy use away from developed and toward developing countries in terms of the growth of the 'consumer class' in much of the developing world (Rubin 2009). This group includes those whose annual income is at least U.S. $7000/year. It currently contains about 2 billion people (almost all of them in the developed world) but the BRIC countries alone—Brazil, Russia, India, and China—will have a population of around 3.2 billion by 2050. With economic growth rates between ten and twenty times those of developed countries, the BRIC countries are adding members to the consumer class at an exponential rate. So global consumption patterns, coupled with the dominance of fossil fuels in the energy mix until at least 2035 (by which point it will be too late to avoid serious climate change), make the socio-economic assumptions underlying the worst-case scenario models look plausible. Again, this path is not foreordained, but avoiding it is going to mean, among other things, that the huge and growing energy demand coming from the developing world is met largely through renewables.

In light of these data, especially the claim by the *Copenhagen Diagnosis* researchers that the already alarming trends we see in the global climate 'will likely accelerate' without significant constraint of the global fossil fuel supply, our refusal to be frankly alarmist about the situation we are in—to talk openly about the possibility of catastrophe and even apocalypse—is hindering us from dealing intelligently with the problem.[3] This might sound like an odd thing to say. Isn't our culture full of references to the imminent catastrophe of climate change? Yes, but not in the right places. Two key groups need to change the way they talk about climate change: mainstream economists and climate scientists. Given the esteem in which they are held, if these two groups spoke less arrogantly (in the case of economists) and less cautiously (in the case of scientists) there would be much greater general cultural pressure to change our policy. Let's look at the two groups separately.

There are two broad problems with the approach of mainstream economics to climate change, and especially the potential for future catastrophe. The first is that economists have a very difficult time accounting for the effects of abrupt climate change in their models. As a result, they cling to the now outdated paradigm of gradual change and tend to underestimate *the very possibility of catastrophe*. The gradualist paradigm assumes that there is a steady correlation along two dimensions: (a) between GHG forcings and concomitant temperature rises; and (b) between the latter and observed effects in the Earth's physical systems as well as human systems. For every fixed quantity of Gigatonnes of carbon emitted we can expect a fixed rise in global mean temperature, and this will, in turn, affect specific systems in fixed ways.

But this paradigm has been debunked by Wally Broecker's dramatic discovery in the paleoclimatic record of abrupt change, radical climate switches taking place sometimes on decadal time scales. Broecker's (2000) pioneering work has helped uncover 'a hitherto unknown characteristic of the Earth's climate system, namely, its ability to make abrupt jumps from one state of operation to another'. (137) The trigger for these events is the crossing of various system thresholds. Observations indicate that this is currently happening throughout the Earth system. For example, although there was a temperature increase in northern Canada of nearly 2°C in the 1990s, the melting of the permafrost had been slow (though steady) because of the large latent heat—defined as the amount of heat calories needed to convert a given amount of ice to liquid water—contained in that mass of frozen matter. But the melting has suddenly started to accelerate rapidly (Steffen et al. 2004, 13–14). No smooth curve here, *pace* the gradualist paradigm. The climate economist's job is to model the costs associated with these kinds of changes and thus to provide policymakers with a rational basis for future-directed decision-making. But these switches are notoriously—perhaps essentially—difficult to predict and they therefore give the lie to future climate-impact predictions based on cost–benefit analysis. Some economists see this clearly. According to Weitzman, for example, because of the possibility of iterated unforeseen catastrophes the future will be marked by 'deep structural uncertainty'. He therefore argues that, 'climate change economists can help most by *not* presenting a cost–benefit estimate . . . as if it is accurate and objective' (2009, 18).

The second broad problem with climate economics is that, even if cost–benefit modelling were not stymied by the abrupt paradigm, and economists were in principle therefore open to the possibility of catastrophe, they misunderstand *what catastrophe means*. The *Stern Review*, for example, does not shy away from talking about the possibility of catastrophe. Stern has argued that if we continue to do nothing about it, climate change could cost the world 20 per cent of its annual GDP by the end of the century, and he certainly seems to think of this as a catastrophic outcome. But this way of putting the point nevertheless under-describes the problem (Dumanoski 2011, 73). Even if the number is correct in some abstract sense, the assumption behind it is that we *can* fully cope with our problems just by diverting our wealth to the appropriate places. It's a lot of money, to be sure, but once we have done this, there will be no remaining costs. But compounding costs this way does not even scratch the surface of the manifold harms and horrors that climate change catastrophes will bring us (a point to which I return in Section 3.5, below). Not all losses can be monetized, a fact that should make us question the economist's ability to tell us very much about what is at stake with climate change.

This brings us to climate scientists. Strong evidence is now emerging that, just like mainstream economists, many mainstream climate scientists are hooked on gradualism. Just as there are exceptions among economists to the criticisms I have advanced about the discipline as a whole, so too there are plenty of scientists who do not fit the description of the discipline of climate science I'm about to offer. In particular, since climate scientists like Broecker are responsible for making us aware of the abrupt paradigm in the first place they can hardly be charged with making the same mistake as the economists on this score. Still, if the more general problem here is what we might call a conservative bias, too many climate scientists display it too. Conservatism is not the view that the world cannot change but that change always happens in accordance with established patterns. This kind of conservatism works hand in glove with gradualism. A salient example of this phenomenon concerns IPCC reports. Contrary to the manner in which they are sometimes portrayed in the media, these reports are deeply conservative documents. In large part, this is a function of the consensus-based model operative in their production. In virtually the only portions read by the broader public—the summaries for policymakers—even governments have a say in the final wording of what are, after all, supposed to be reports of purely scientific findings.

As a result of these and other pressures, IPCC reports have consistently underestimated key developments in natural systems, such as the rate of arctic sea-ice decline (Allison et al. 2009), extent of sea-level rise (Rahmstorf et al. 2012), rate of CO_2 emissions, especially in China (National Academy of Sciences 2009), loss of Northern Hemisphere snow cover (Déry and Ross (2007), and rate of permafrost melt (United Nations Environment Program 2011).[4] There is more than one reason for these failures of prediction. Take the example of Arctic ice, which is thinning four times as quickly and drifting twice as quickly as predicted in the IPCC's 2007 report. The problem is that the IPCC based its prediction—that the ice would not melt until 2100—on temperature fluctuations alone, whereas new research suggests that proper modelling should include underlying mechanical forces like wind and ocean currents which tug violently at the ice, undermining its integrity (Romm 2011). Something similar can be said about the other inaccuracies, many of which have to do with a perceived inability on the part of the IPCC to model feedbacks accurately. These kinds of error and the corrections to them that hopefully ensue are a matter of course in the progress of science. Indeed, I will argue in Chapter 5 that in spite of errors like this, we non-experts are qualifiedly justified in believing what IPCC reports are telling us.

Here, I want to focus instead on a different source of 'error': that involved in misrepresenting dramatic discoveries in the communication of scientific results. As it happens, too many climate scientists have been 'erring on the side of least drama' (Brysse et al. 2013).[5] The phenomenon is so widespread it has acquired an abbreviation: ESLD. The findings are surprising and important because in the popular imagination climate scientists often have a reputation for exaggeration. But as Brysse et al. have argued, most 'scientists are biased not toward alarmism but rather the reverse: toward cautious estimates' (2013, 327). Indeed Hansen argues that, as a group, climate scientists have an unfortunate 'tendency for gradualism' when faced with new and startling evidence. He claims that this has prevented them from grasping, or at least 'effectively communicating', the magnitude of the threat posed to the Greenland and West Antarctic ice sheets, for instance (Quoted in Brysse et al. 2013, 330). By way of contrast, note the reaction of James Lovelock (in a letter to a friend) to the discovery of disappearing ice in the Arctic:

> Have you seen this week's *Nature*? There is an account of the rapid melting of Arctic floating ice and the author says that in twenty years the Arctic ocean will be ice free in the summer. This is *truly apocalyptic*, yet they still see it as a regional not global problem. (Quoted in John and Mary Gribbin 2009, 206; my emphasis)

Neither Hansen nor Brysse et al. are particularly critical of scientists for taking the cautious stance to which Lovelock is pointing here. Scientists are afraid of making dramatic pronouncements because they are so often caricatured. It's very easy to dismiss the messengers as Jeremiahs. Scientists are framing their communications in cautious and gradualist language partly because they want to be heard (and they believe this is the only way to do so) and partly because to some extent they themselves doubtless share the relevant assumptions about the future.

So ESLD, according to these authors, is merely the regrettable upshot of otherwise praiseworthy scientific values like 'rationality, dispassion and self-restraint' (Brysse et al. 2013, 328). This response to the phenomenon is inadequate because when such values get in the way of describing and responding to risk with an appropriate sense of urgency, they are not praiseworthy. Rationality, it hardly needs to be said, does not demand that we sugarcoat the facts just to get them heard, especially when this distorts our policy responses. Dispassion can be misplaced. If my child is drowning in a pool, am I at all aided in doing the right thing by approaching the emergency dispassionately? As for self-restraint, it can be a motivational enabler of the bystander effect. Seeing a person being assaulted, I might think I'd do best to restrain myself from confronting the attacker since that burly spectator over there is clearly better suited to intervene.

We do not have time for this sort of prevarication nor for the constitutional myopia of the dismal science. Recall the notion of temporal stretching from Chapter 2. There it was argued that since our activities have effects well into the future, the interests of future people are implicated in what we do right now. Similarly, we should think of the present as it is currently structured as *embedding* a specific future. The catastrophic world our practices are leading to is *an aspect of present reality*, one no less real than those practices themselves. This is why I have been insisting so strenuously on the need to make all of this explicit. Jameson says that a good way to understand the Marxian concept of 'reification' in the age of globalization is that it represents the 'effacement of the traces of production' from the objects we consume (1992, 314). The point is that the traces are evidence of past, and quite likely persisting, social conditions that make claims of justice on us.

But there is no reason for this sort of analysis to be confined to the way in which presently available products bear the traces of their past production. Moving in the other temporal direction it might be said that we reify our practices as well as the system that supports them to the extent that we efface from them any connection to the world they are bringing into being. That connection also makes demands of justice on us. So let's be blunt in saying that we are squarely on a path to climate catastrophe by the end of this century. The 5°C world to which we are likely headed is one the planet has not seen for millions of years, perhaps not since the Eocene period some 55 million years ago (Ward 2007, 167). In Mark Lynas's summary of the models, with just a 4°C rise, the rainforests are burning away, the great ice sheets of Greenland and West Antarctic are in a death spiral, many of the world's huge coastal cities are inundated, severe and prolonged drought grips much of the equatorial belt of the planet, ecosystems and biodiversity are under severe assault, super storms—like Typhoon Haiyan which tore through the Philippines in November 2013—proliferate and agricultural productivity plummets (Lynas 2008, 163–90). Add yet another degree Celsius and not only do these effects become more pronounced, we will also see a huge outward expansion of the world's deserts, a major decrease in the flow of rivers like the Nile, the near total loss of winter snowpack in places like California, the drying up of underground aquifers, soil desiccation, increasingly ferocious wildfires, and more. And, finally, at this level of heat increase the collapse of global agriculture extends even to northern Canada and Siberia (Lynas 2008, 193–214).

This is not a picture of the future in the abstract. It *is* business as usual. It is the future-as-contained-in-the-present. Things look very bad but we should not conclude that there's nothing we can do to avoid climate catastrophe. This is to indulge in the false and self-serving practical pessimism against which I argued in Chapter 2. My claim here is that it is imperative for the catastrophic framing mode to move from the environmentalist fringe to the cultural mainstream, so that speaking and thinking this way becomes the general social habit. That's the point of focusing my critique on scientists and economists. Again, these are two of the most prestigious groups in our cultures. If they could be convinced to speak more forthrightly about the dangers we face, and about the challenges these dangers pose to the method-ologies and presuppositions of their respective disciplines, the rest of our culture—including our politicians—would, I think, swiftly follow their lead.

I don't want to overstate the point here: appeals to the imminence of catas-trophe can't do all the motivational work. Research in this area shows that everything depends on the form of communication, with an emphasis on the distinction between two information-processing systems: the cognitive and the affective.[6] If our communication of the facts is the dry, abstract, and statistical sort favoured by scientists, people will tend not to be engaged by it to the extent of acting so as to help avert the negative possibilities we are trying to highlight. This is especially relevant for climate change. After all, the concept of climate is purely statistical, and much of the data with which scientists have confronted us concerns the way the world will be in the relatively distant future. On the other hand, people do get motivated to act when their emotions are also engaged, but not if the appeal to fear is predominant. So we appear to be in something of a bind with respect to communicating the threats posed by climate change. If we merely cite statistics about temperature anomalies, sea-level rise, loss of biodiversity, changing isotherms, or even likely death rates (and so on) we will be met with a shrug. If we dramatize these risks in disaster films and the like we will scare people out of their wits. Either way we succeed only in numbing them rather than mobilizing them. These are empirical claims about the way people respond to the communication of disaster or risk. Let me make three points in response to them.

The first point highlights an important corollary of the Kantian doctrine that ought implies can, namely that can implies the possibility of ought. As long as we can act in accordance with appropriate moral standards in some situation no empirical claim concerning the way we tend to behave in such situations can determine the scope of our duties there. This is a book about the moral challenges of the climate crisis, challenges centred on the need to cultivate three virtues: truthfulness, justice, and hope. Thus I claim (in Chapter 5) that the virtue of truthfulness requires that we look this crisis in the eyes. In my view, there has to date been far too much evasion, temporizing, and outright lying about the issue. So if it is true that we are on a path to catastrophe, this should be stated in as unequivocal a way as possible. In the interests of accuracy it is not possible to avoid heavy reliance on statistics here. Coming to moral requirements, the virtue of justice (examined in Chapter 4) tells us that if we have a duty to do our bit to help avert catastrophe then, so long as catastrophe is actually avoidable (which I think it is), that's what we should do. Finally, I argue (in Chapter 6) that given the importance of the issue we have a duty to hope for a good outcome to this crisis. In sum, so long as appropriate actions are available to us, it matters little that many of us do not respond properly to the abstract communication of morally bad outcomes, provided we understand the communication.

However, and this is the second point, moral philosophy would be a stunted enterprise indeed were we to confine it, as we too often do, to the abstract specification of requirements (as well as the facts supporting them). So I agree that our affective capacities need to be engaged in this communication. The question how to motivate people to change therefore constitutes a major theme of the chapters to follow. I have character sketches in Chapters 4 and 5 that are meant to show us how this works in the twin cases of moral weakness and self-deception. Moreover I have opted for a virtue-focused approach to climate change in the first place because, as I have said, it is a finely tuned theoretical tool for understanding moral agency. For example, my analysis of justice contrasts sharply with the bulk of justice-oriented climate ethics because as a rule the latter treat our moral commitments in a manner that is, in Paul Woodruff's phrase, merely 'mind deep' (2011, 165). Justice is crucial, but when we conceive it as a virtue we are forced to examine the way its proper exercise engages shame, honour, anger, courage, and more. I tell a similar story about truthfulness and hope. As I hinted in Chapter 1,

I show that each of the three virtues needs the assistance of specific auxiliary virtues and emotions if it is to do its moral-psychological job properly.

Finally, it is surely correct to say that fear cannot do all the motivational work here. Left to its own devices it probably will simply numb us. So although it has a significant role to play in my analysis—I won't apologize for trying to scare the hell out of my readers in this chapter as well as parts of Chapter 6—I also want to emphasize and foreshadow the positive aspects of my analysis. For instance, it is not accidental that the book's main argument ends with a substantive treatment of hope, an essentially forward-looking, positive, and *energizing* virtue. More generally, in focusing on justice, truthfulness, and hope as virtues I am trying to describe what many of us want to be. These are aspirational virtues. So there's some light at the end of this tunnel; but first, more of the darkness.

3.3 The New Tribalism

Anthropogenic climate change is causing, and will increasingly cause, a broad range of harms to non-human nature. Biodiversity will likely plummet over the next century and beyond, for example. This is a clear moral failing on our part, one that is indicative of a lack of respect for something we should respect—nature.[7] However, since I have argued in Chapter 2 that the cosmopolitan ideal (suitably refined) is *the* moral ideal of the Anthropocene age, and because this ideal prioritizes what we humans owe to each other, I will concentrate in the remainder of this chapter on a particular failure with respect to this ideal. I have in mind the role climate change is going to play in the rise of violent inter-group conflict. According to Vogel, '[t]here is every reason to believe that as the 21st-century unfolds, the security story will be bound together with climate change' (Quoted in Lee 2011, 1). Among experts on the security threats of climate change there is increasing awareness of the link between climate change and the kinds of security issues that can produce armed conflict among states as well as sub-national violence among ethnoreligious groups (Moran 2011). In the world's most vulnerable regions, some of which are discussed just below, inter-group conflicts have been stoked for many decades by already existing environmental stressors such as the degradation of agricultural land, deforestation, depletion of freshwater resources, depletion of ocean fisheries, and desertification.

The causal chains are, however, complex. The environmental stressors just listed are, taken alone, not usually sufficient to cause large-scale violence. But when combined with population growth and unequal access to resources, they can produce significant resource scarcity in a society. This, in turn, can have two effects: migration and/or expulsion of large numbers of people and decreased economic productivity. Migration and expulsion often lead directly to ethnic conflict, when for example migrants push out or otherwise challenge the hegemony of native inhabitants of the lands they move into. But both migration/expulsion and scarcity-induced decreases in economic productivity can also lead indirectly to group violence via the weakening of the state (Homer-Dixon 1994, 31). To make matters even more complex, there can ensue a feedback effect such that violent conflict exacerbates the original environmental degradation that caused it (Brown et al. 2007, 1148). Sometimes the only alternative to full-out civil and political fragmentation is an increase in authoritarianism, a solution that obviously carries its own dangers for human security.

The following claim by Thomas Homer-Dixon about the effects of environmental degradation on human security could, *mutatis mutandis*, apply to all of the countries we are about to examine (and many more besides):

> Countries experiencing chronic internal conflict because of environmental stress will probably either fragment or become more authoritarian. Fragmenting countries will be the source of large out-migrations, and they will be unable to effectively negotiate or implement international agreements on security, trade and environmental protection. Authoritarian regimes may be inclined to launch attacks against other countries to divert popular attention from internal stresses. (1994, 40)

This sounds a lot like what went on in the seventeenth century. Future climate-induced violent conflict will be concentrated in two geographical zones, what Lee calls 'tension belts'. The most pressing set of worries is focused on a wide band of countries extending at least 35° latitude north and south of the equator, and in some cases even slightly further than this. Here we will likely see increased violence as a result of processes of desertification, loss of freshwater resources, and inundation of coastlines.[8] In the Western Hemisphere, this region comprises the northern part of South America as well as Central America, Mexico, and the American Southwest. In Africa, the area stretches from Morocco to Somalia, then through much of the Middle East and a wide swath of Central and South Asia.

In all, more than forty potentially problematic countries are in this region. Since billions of people live there it is useful to look more carefully at a few potential hotspots. The selection of countries is, to some extent, arbitrary, but not entirely. I have attempted to sketch the potential for violent conflict in places that have large populations and that, collectively, span a good part of the globe. More could have been added, for example Vietnam, The Philippines, Pakistan, Egypt, Southern Africa, The Maghreb, Brazil, and the southern United States. But the story is more or less the same wherever we look: as the impacts of climate change manifest themselves in the coming decades a clear pattern of increasing violence among and, especially, within states emerges.

Let's begin with China, where climate change has historically been a key factor in determining the outcomes of dynastic struggles as well as periodic occurrences of war and general civil strife (Lewis 2011, 12). One of the major dangers for China in a warming world is associated with rising seas. In China, about 144,000 square kilometres of land is 5 metres above sea level or less. These lands are for the most part set in the deltas of the Yellow, Yangtze, and Pearl Rivers (Lewis 2011, 14). If there is significant flooding in these areas, we could see huge migrations of people from the south, especially from densely populated urban centres on both the Yellow Sea and the East China Sea—Shanghai, Ningbo, Linhai, Tianjin, and Qingdao.[9] In the short-term, these refugees will probably move to cities further to the west—Beijing, Lanzhou, and Zhengzhou, among others. But this could cause immense problems because China's distribution of resources is a veritable powder keg. The cities in the east are the economic powerhouses, but they contain few minorities (this is generally a recipe for intense strife as ethnically diverse outsiders move in); the south-east and central eastern regions are primarily agricultural, but lack abundant water resources, and the west possess significant water resources (Lewis 2011, 16). There is also a real worry about climate refugees from the north. Both Mongolia and northern China face the prospect of severe desertification by 2030. As Lewis notes, this will both worsen the problem of water scarcity and lead to influxes of migrants from Inner Mongolia and Tibet, which 'will increase the risk of ethnic conflict in areas that currently lack such minorities' (Lewis 2011, 15–16).

Declining water supplies on the Tibetan Plateau will directly affect the vital interests of at least 500 million people in Asia, half of them Chinese. As Homer-Dixon argues, there could be 'bitter disputes' among people in these regions, in the form of 'deprivation and identity conflicts' (1994, 38). According to Goldstone, 'China may once again, as it has so often in its history following the fall of unifying dynasties, experience a decade or even century-long interregnum of warring among regional states' (Quoted in Homer-Dixon 1994, 39). Finally, there is the possibility that the Chinese might act aggressively to protect water resources for its people by diverting rivers with headwaters on the Plateau from countries such as Vietnam, Myanmar, Laos, Pakistan, Nepal, and India (Lewis 2011, 17). This could lead to war with some of these countries, two of which possess nuclear weapons.

The second country to look at is India. In India, climate change is, according to T.V. Paul, likely to 'accentuate preexisting conflict patterns...More generally, however, climate change could exacerbate the problems associated with state weakness in multiple areas, including the preservation of social and political order' (2011, 73–4). The main threats here have to do with the interruption of the monsoons, the regularity of which is crucial to Indian food production. A significant drop in agricultural production would be disastrous for hundreds of millions of Indians. One effect of this would be to widen the already significant gaps between rich and poor as well as urban and rural, causing many more Indians to be 'ensnared by the "poverty trap"' (Paul 2011, 76).

Because many of the countries surrounding India will probably be in even worse shape than it is from the effects of climate change, there are likely to be fewer problems associated with out-migration and more with internal move- ment, especially the movement of large numbers of people from the country to the cities (Paul 2011, 76–7). One possible outcome of this is an increase in the power of the Naxalite movement throughout many Indian states, especially those stretching from the border with Nepal to Andra Pradesh. For years, the Naxalites have capitalized on the grievances of the poor and landless to wage guerilla war against the central and state governments. Their appeal, and sway, is likely to increase dramatically with increased climate-induced impoverishment. And more generally, according to Paul, because economic development has been so uneven in India, and because climate change looks as though it will exacerbate this unevenness, violent insurgency movements in Nagaland, Mizoram, Manipur, Tripura, and Assam could proliferate. At the moment there are over one hundred insurgency movements in these regions (2011, 76–7).

Moreover, such conflict would not have its sole source in economic alien- ation but also in pre-existing divisions based on caste and religion (Paul 2011, 77). Since India remains a relatively strong state, none of this is likely to lead to significant political fragmentation, at least not in the short term. But this only forces us to the other horn of Homer-Dixon's dilemma: to cope with all this internal unrest the central government may become much more authoritar- ian. Moreover, the exacerbation of internal divisions along the lines indicated will almost certainly curb India's push for economic growth, a fact which could create a kind of social feedback as perceived stagnation widens social divisions even further. The potential for violence along these fault lines thus becomes more pronounced (Paul 2011, 75–7).

Another disturbing possibility for India has to do with its relations with some of its neighbours. The main threat here comes from Bangladeshi climate refugees. Bangladesh's mean elevations vary from less than 1 metre on the tidal floodplains—like the Sundarbans on the south-west coast—to more than 30 metres in the north-west. A sea-level rise of just 1 metre in Bangladesh would cause the loss of 17 per cent of its cropland, causing up to 10 million people to become landless (Riaz 2011, 104). Since many millions more are indirectly dependent on the agricultural and fisheries output of the Sundarbans region, widespread inability to meet basic subsistence needs is a likely out- come. All of this will compound, and be compounded by, losses that were set to hit the country even without the effects of climate change. Population increase alone will have a devastating effect on subsistence: Bangladesh's population is predicted to reach 235 million by 2025, thus reducing by half the amount of cropland available per capita (Homer-Dixon 1994, 21). Bangladesh is already one of the world's poorest countries so there is no chance its government could cope with this. Although many of the refugees would flee to Dhaka—already overcrowded and 'politically volatile' (Riaz 2011, 106)—many will also attempt to migrate north to India.

Indeed, this has already begun, and the way it has transpired so far gives one little hope that increased strife can be avoided. Over the years, as Homer-Dixon has noted, the movement of people from Bangladesh to India has 'triggered serious intergroup conflict'. For example, the Lalung tribe in Assam has accused Bengali Muslim migrants of usurping the best farmland. In 1983, 1700 Bengalis were killed in a 'five hour rampage' by members of the Lalung tribe. And in Tripura, throughout the 1980s there was continuous armed conflict between the state's original Christian and Buddhist inhabitants and Hindu migrants from Bangladesh and Pakistan (Homer-Dixon 1994, 23). Although the government eventually made some effort to halt the flow of Bangladeshi migrants, thus ameliorating the conflicts, climate change will, as we have seen, place already precarious settlements like this in serious jeopardy as a result of what may be an uncontrollable flow of refugees.

Consider, third, the status of some West African countries—Côte d'Ivoire, Nigeria, and Senegal—bordering or close to the Sahel, a semi-arid zone running across northern Africa south of the Sahara and north of the tropical savannah to the south. In this region flooding and droughts are the main problem. The German Advisory Council on Global Change has recently concluded that climate change in this region 'will lead to increasing regional destabilization, including increased potential for violent conflict' (Quoted in Beck and Pires 2011, 204). Côte d'Ivoire is already beset by significant ethno-religious conflict and there is some indication that government officials actively 'encourage a culture of violent xenophobia' for political gain (Beck and Pires 2011, 206). For instance, it has been reported that supporters of the government regularly incite violence against 'Dioulas', a term for anyone deemed Muslim (Beck and Pires 2011, 206). Climate change could exacerbate these tensions because Côte d'Ivoire is likely to experience grave basic resource shortages over the next twenty years and beyond. According to the Nairobi-based International Livestock Research Institute, by 2050 Côte d'Ivoire will see declines of more than 20 per cent in the length of its growing season in many of its agro-ecological zones. All of this will happen in regions that have historically experienced resource conflicts and violence, for example between Peuhl pastoralists and farmers in recent times (Beck and Pires 2011, 207).

Two further threat multipliers should be mentioned in the case of Côte D'Ivoire. The first is that as a result of further population growth and increasing urbanization (among other things), Côte d'Ivoire is also likely to face severe water shortages in the coming decades: the percentage of Ivorians facing freshwater scarcity (defined as less than 1000 cubic metres per capita per year) will increase from 0 per cent in 2000 to a staggering 23 per cent in 2030 (Beck and Pires 2011, 207–8). The second is the likely influx of climate refugees from neighbouring countries. We have already seen that such flows are often the fuel that ignites pre-existing ethno-religious tensions. As Beck and Pires note, West Africa has 'a long history of migration occasioned by environmental, political and economic stress', a phenomenon likely to increase as a result of climate change (2011, 208).

Nigeria is likely to see climate-induced ethno-religious tensions in its northern Sahelian region as well as problems along the heavily populated southern coast stemming from rising sea levels. The north is a semi-arid region whose agricultural output will almost certainly decline because of reduced rainfall. This will very likely sharpen already-existing tensions between the Muslim north and the Christian south, as well as putting fully one-quarter of the population into a situation of water scarcity (Beck and Pires 2011, 211). Although southern cities like Lagos will experience an influx of internal migrants, the north will be hit especially hard. Many will attempt to move south, but given the historical animosities as well as the fact that southerners will be coping with the effects of rising seas they will likely not be welcomed with open arms.

Because much of the country is in the ecologically troubled Sahel, the effects of climate change will be worse in Senegal than in Côte D'Ivoire or Nigeria, even though Senegal, in view of the substantial power the constitution accords to its central government, is probably better able to cope with these problems than the others are (Agbor et al. 2012). In Senegal, a key source of socio-economic tension has been conflicts between farmers and pastoralists, especially down the length of the Senegal River: '[a]lthough these conflicts are often among groups of Senegalese, the worst violence over land use took place in 1989 between Senegalese farmers and Mauritanian pastoralists and flared into massive human rights abuses in both countries' (Beck and Pires 2011, 214). As with many Sahelian countries, the main issue for Senegal has to do with access to freshwater resources. By 2030, nearly three-quarters of the population of that country is expected to face water scarcity. Inability to irrigate sufficiently will cause a massive loss in agricultural output: up to eight times the global rate of decline by the end of the century (Beck and Pires 2011, 215–16). Finally, Senegal has a long coastline supporting a huge population—nearly three-quarters of the country's population lives within 100 kilometres of the coast (Beck and Pires 2011, 216)—whose ability to survive will be severely challenged as seas rise and urban populations swell.

3.4 Discontinuity and Crimes of Atrocity

The situation with these African countries is instructive on a number of levels. For one thing, we should resist positing deterministic causal links between climate change and violent conflict. This is not just because, as we have seen, the causal chains are circuitous, but also that there are countervailing forces. As researchers have noted, African countries have been dealing with adverse environmental conditions for centuries and they have learned how to avoid total social collapse through such strategies as diversifying the means of securing a livelihood, altering norms of governance, migration, changing agricultural practices, and so on (Brown et al. 2007, 1149). However, the key question is whether or not every country can rely on strategies like this to deal adequately with rapid, unexpected changes in the basic environmental conditions associated with abrupt climate change.

As indicated, climate change is likely to accelerate the key causal pathways between climate change and violent conflict, chiefly by intensifying some of the original environmental and social stressors. However, as Lee has pointed out, such conflicts will emerge only after a 'sustained period of divergent climate patterns' (2011, 4). It has until recently been difficult to be precise about when this divergence will occur, either as a global average or on a regional or national basis. New research is providing a clearer picture of this, however. One way to understand the problem is by reference to the concept of 'stationarity'. This refers to the ebb and flow of natural systems—freshwater, fisheries, forests, and so on—within understood parameters. For most of these systems, past variability of the resource was not so great as to preclude 'stationarity-based design'. This allows for intelligent management of the resource over time. One of the variables kept relatively constant for this purpose has been climate. For instance, relative historical climate stability has allowed for significant investment in, and development of, water infrastructure, which up until now has exceeded U.S. $500 billion annually (Milly et al. 2008).

However, scientists now claim that all this is under threat from climate change and that as a result 'stationarity is dead' (Milly et al. 2008, 573). This may be an overstatement, but it does appear as though the concept is at least in its death throes and that climate change is the main culprit in its imminent demise. A recent scientific report tried to pinpoint when we might expect a 'radically different climate' to take hold. Two pathways were modelled, one involving 'concerted rapid CO_2 mitigation', the other a business-as-usual scenario based on RCP 8.5. Scientists quantified minimum and maximum surface temperature values from 1860 to 2005 (or in some cases 2010) to establish the stationarity envelope for each region of the planet as well as the oceans. Not surprisingly, the year at which projected temperatures move consistently outside the envelope is different for the two pathways.

With concerted effort at mitigation, the global average year is 2069 (2072 for the ocean), while for business-as-usual it is 2047 (2051 for the ocean). Of course, the global average hides significant regional variability. For our purposes, the key is that equatorial and tropical zones will be outside their historical variability envelope as early as 2015–2020 (2034 for the ocean in this region). This will have a huge impact on the Earth's biodiversity hotpots, the rainforests and corals. As for human impacts, the authors of the study point out that up to 5 billion people live in places whose climate will be in a state of perpetual discontinuity by 2050 (Mora et al. 2013).

From the perspective of possibilities for social adaptation what in general terms does the threat of radical climatic discontinuity represent? Putting together the arguments I have made about our current pathway and the empirical research on climate change and violent conflict, we appear to be securely on the road to accelerating social and political chaos in much of the world. We need to emphasize the point about *perpetual* discontinuity. We are not talking about a single break with the past, after which a new period of stability establishes itself. That would be relatively easy to adapt to. Instead, surprises and breaks with expectations based on past patterns would appear every year, perhaps every month. The adaptation measure that worked last year or last month might suddenly be impotent in the face of surprising new challenges. That levee we built after last year's storm surge and which we thought would be sufficient for many years to come proves to be insufficient for this year's challenges. How does a government, or a people, cope with perpetual climatic novelty?

Nor is this simply an engineering problem. Because the Anthropocene, as we have seen, is marked by feedback loops between the natural and conventional worlds, governments must also think about the ways in which adaptation measures might affect the cohesion of human communities in both the present and the future. As Simon Dalby has argued, 'consideration of ways of adapting without inducing conflict is now part of the environmental-security agenda' (2013, 38). Here is how the problem is described in AR5:

> The capacity for collective action is a critical determinant of the capacity to adapt to climate impacts, and this too can be undermined by violent conflict... When conflict exacerbates existing horizontal inequalities between ethnic or religious groups, foments distrust in local or government institutions, or isolates individuals and households, the social capital that is important for adapting to climate change is also degraded. Conflict-related displacement also disrupts social networks and makes it difficult to achieve elements of secure livelihoods, such as marriage, access to land, or access to communal social safety nets. In situations of violent conflict, efforts to address climate change that provide financial or resource flows that can be captured by local elites or illegitimate institutions, may compound divisions and exacerbate grievances. (IPCC 2014a, 774–5)

Security experts have concluded that in the age of climate disruption human insecurity stems from three main problems. First, 'democratic deficits', understood as significant socio-economic inequalities. Second, adaptation measures that exacerbate existing tensions, whether or not they are intended to do so. Third, 'structural violence', understood as 'institutionally sanctioned practices and decision-making processes' that contribute to the exploitation of marginalized groups (Zografos et al. 2014). If these problems are not addressed robustly as societies adapt to climate disruption, violent conflict could escalate to the point at which it transforms into state-sponsored crimes of atrocity. As the examples I have sketched indicate, in many places in the world, climate change will increase fragmentation among people along religious, ethnic, class, and geographical lines. As we have seen, in the region we have examined, one or more of these three problems is pronounced.

This means that those who live there, or in other vulnerable places, will experience lives of increasing uncertainty and stress, the ultimate manifestation of which is a constant state of Hobbesian war: not only, or even primarily, actual fighting but also 'the known disposition thereto' (Hobbes 1994, part I, ch. xiii, 76). For example, a recent study of a Bosnian village showed how a weakened state, coupled with economic contraction, led a previously pluralistic community into increasing ethnic fragmentation between 'two groups who respectively dehumanized, and ultimately began killing each other' (Barnett and Adger 2007). Ultimately, social and political forces like those operative in the Bosnian case can cause a tense but controlled situation to devolve into mass violence—culminating in genocide—and this is an outcome we cannot discount for climate-induced conflicts.

Genocide expert Ervin Staub has argued that 'difficult life conditions' are often key contributors to genocide (Quoted in Alvarez 2009, 130). There is evidence that in tough times like these, states will often turn to repressive policies, and that these policies often involve targeting a subgroup in society. More importantly, such measures are reliably accompanied by a general willingness among members of the favoured group to support harsh treatment of members of the disfavoured group (Costelloe et al. 2007).[10] This sort of scapegoating does not necessarily lead to specifically genocidal forms of violence, but it can. One of the key lessons from social psychology in the past thirty years or so is that crimes of atrocity do not require the agency of demons. Here is how James Waller puts the point:

> As we look at the perpetrators of genocide and mass killing, we need no longer ask who these people are. We know who they are. They are you and I. There is now a more urgent question to ask: *How* are ordinary people, like you and me, transformed into perpetrators of genocide and mass killing? (2007, 137)

Waller shows that 'ordinary' people—the respectable sorts you thought you knew well—are capable of terrible acts of violence if conditions allow for it.

He focuses on purely social conditions like the cultural construction of a worldview stressing authority orientations and social dominance; the psychological construction of the other through Us–Them thinking, 'moral disengagement' and victim blaming; and the social construction of cruelty through various forms of group identification (2007, 138).

My claim is that these factors will be given much greater expressive scope by the radical climatic discontinuity that is coming in the decades ahead. One possibility is that in extreme cases—and by hypothesis more and more events will fit this description—the state will run out of material resources to combat the latest surprise manifestation of climate change. Lacking better ideas and seeking to preserve itself, it may scapegoat a subgroup. This points to the ambivalent role of the state in combating the effects of abrupt climate change. We have seen that a strong state is required to provide reliable adaptation but, again, if the state fails to do this and chaos looms, state officials might be tempted to encourage group violence. This could have the effect of diverting attention away from governmental failure.

For example, states are very adept at reinforcing social identity by reminding citizens of historical grievances and identifying enemies. To illustrate this, come back to a future China's potentially flooded coastlines. As we have seen, the short-term solution to the inundation of the southern coastline will likely involve the movement of people into Chinese cities to the west. But as these cities begin to feel the stress of this huge influx, a more extreme measure might be contemplated. For a vivid example of what could happen, consider the scenario described by James Woolsey:

> Northern Eurasian stability could...be substantially affected by China's need to resettle many tens, even hundreds of millions from its flooding southern coasts. China has never recognized many of the Czarist appropriations of Chinese territory, and Siberia may be more agriculturally productive after a 5–6° C rise in temperature, adding another attractive feature to a region rich in oil, gas and minerals. A small Russian population might have substantial difficulty preventing China from asserting control over much of Siberia and the Russian Far East. The probability of conflict between two destabilized nuclear powers would seem high. (Quoted in Dyer 2008, 61)

As indicated, were something like this to happen, the Chinese government would very likely attempt to justify its actions, in part, by appeal to what it takes to be a specific historical injustice: nineteenth-century appropriations of formerly Chinese territory by the Russians. This is important because it could provide a sense of historical mission to the Chinese leadership—and through it to the people—making it easier to target Russians in the Far East as usurpers of territory that is Chinese 'by right'. As Alvarez has noted, perceived past victimization is a powerful facilitator of crimes of atrocity (2009, 64). Moreover, once the violence has begun it is very difficult to turn it off, and it can feed back into, and thereby exacerbate, the very environmental problems that helped cause it.

Finally, it's worth pointing out that all of the countries I have been examining here were studied to only 2030. In every case, the prospect involves a steady increase in inter-group violence, but none of the authors focusing on these security problems takes account of the new scientific research projecting dates for a 'radically different climate' (the latter was published two years after the former, so this is not a criticism). Putting these two bodies of research together allows us to predict that inter-group violence in these and other areas is likely to increase dramatically beyond 2030. The result is that we cannot discount the possibility that there will be significant genocidal violence in the twenty-first-century. This is a world coming apart at the moral seams.

3.5 Reconceiving Adaptation and Mitigation

Clearly, it makes a good deal of intuitive sense to talk about this emerging world as representing a threat to security. But the traditional conceptual focus of security concerns—i.e., 'national security'—is weighted with problematic moral and political associations so we need to be cautious using it. For our purposes, the shortcomings of the concept are most evident in a 2007 report by the United Nations Environment Program (GEO4). Using IPCC data about future scenarios, the report sketches four future scenarios, each of which prioritizes a specific goal: markets, policy, security, and sustainability. Dalby summarizes the security first option this way:

> "Security first" focuses on keeping the existing distribution of power and wealth intact, paying attention to the environment only insofar as it is understood as a source of resources for the global economy. It's a narrow view of security, which enhances controls on migration but facilitates the expansion of trade... Protecting habitat gets little attention and coal technology has a resurgence that increases atmospheric pollutants and greenhouse gas emissions. Climate change puts further strain on food production and water availability, and conflicts in Africa in particular are aggravated by these shortages. (2009, 93–4)

This is what happens in the age of climate change when the security emphasis is on the stability of currently existing state structures and the economic arrangements underpinning them. There are three problems with this approach. First, as Deudney warns, it is the sort of framework that invites militarization:

> Environmental degradation is not a threat to national security. Rather, environmentalism is a threat to the conceptual hegemony of state centered national security discourses and institutions. For environmentalists to dress their programs in the blood-soaked garments of the war system betrays their core values and creates confusion about the real tasks at hand. (Quoted in Dalby 2013, 51)

Second, this response powerfully reinforces current political and economic power structures (Dalby 2009, 53). If these structures are part of the problem in the age of climate change—which, as we have seen, they can be—then this is clearly an inadequate approach to security. The third point follows on this one: the approach ignores the root causes of insecurity, which, in Dalby's words, can very often be found in 'poverty and the global disruptions of natural systems' (2009, 50).

Dalby's work on security is important because he shows that the Anthropocene demands an entirely new conceptualization of security from us (the move to 'human security' and 'environmental security' in recent security discourse partly addresses this concern, but not fully). As we have seen, the key feature of life in the Anthropocene is the feedback loops between the human and natural worlds. Now that we have become world-shapers the quest for security can no longer be separated from considerations of what we aspire to become. There is, or should be, an emphasis now on our conscious reality-making activity that is missing in the traditional picture of seeking protection *from* this or that threat. So long as this activity is guided by extended cosmopolitan moral principles rather than science, economics, or technology there is no reason to fear it. Just as a mountain climber seeks a reliable path to the summit rather than security from death or injury, so we need to think of our institutions and the natural world in terms of their *reliability* in securing the goods we need in order to flourish. This indicates that security is at best an instrumental good, and so my plea here is that we deflate it accordingly. We value security only insofar it allows us to achieve other good things, some of which may be intrinsically valuable: love, friendship, reasonable material prosperity, the goods of culture, access to unspoiled nature, meaningful work, and so on.

If citizens have their eyes on the proper ends to which security measures are at best a means, they will be in a better position to see through the illegitimate appeal to force for the alleged purpose of restoring or maintaining security. Put into practice, my deflationary suggestion can act as a check on the proposed use of force to enact this or that security measure. Citizens and bureaucrats can ask whether or not the proposed measures are, measured by past experiences, likely to increase their chances of achieving the things they *really* value. With this modest suggestion in place, let's move to the more important question. On what exactly do we rely as we seek security in our lives? Broadly, the two kinds of things we need to attain the goods just mentioned: our social and political institutions—thought of so as to encompass the way we manage or relate to the rest of the biosphere—and specific other people. Unfortunately, the securitization of climate change, even where human security is emphasized, has focused disproportionately on the first of these, the ways in which individuals' vulnerabilities to large-scale social and political processes or environmental events—war, corruption, economic policies, migrations, natural disasters, and so on—can be minimized through specific political and institutional arrangements (Brown et al. 2007, 1150).

But our reliance on specific other people in the quest to enhance the security of our lives is also crucial, so in what remains of this chapter I want to examine this element more carefully. What we expect from others is that to the extent that we are ourselves reliably, if not infallibly, moral, they will usually repay us in kind. As we will see, in the uncertain times to come this underscores the importance of the virtues considered as stable dispositions of character. A potentially fruitful way of couching the discussion is by reference to the distinction between adaptation and mitigation, both of which are required as we address the complexities of climate change. But whereas duties of mitigation fall chiefly on the prosperous of the world, the demanding task of adaptation will, at least in the short to medium term, fall on the world's poor.[11] In both cases, however, an appeal to the virtues and vices can be illuminating.

Let's begin with the demands of adaptation. In Plato's *Republic*, Glaucon and his friends ask Socrates to give an account of virtuous conduct as an end in itself. What Glaucon calls the 'extreme of injustice' is a situation in which the true state of one's soul is taken by one's group for its opposite. In this scenario, the truly just person has the greatest reputation for injustice and is duly reviled, while the unjust person has the greatest reputation for justice and is duly celebrated (Plato 1992, 361a-e, 36–7). Although Socrates goes on to provide the account his interlocutors are asking for, showing that justice is indeed its own reward, the larger burden of the *Republic* is to prove that truly just people cannot flourish in an unjust world. If the world does not meet them halfway—in the shape of just institutions, other just people and the *reliability* these things impart to social life—such people can never fully develop. The Platonic ideal of justice specifies that justice in the soul and justice in the city are two faces of the same coin. This is what I examined in Chapter 2 as the ideal of non-alienation. Reason seeks an intelligible unity between what agents do for good reasons and the way the rest of the social and political world takes up their doings. This is not to say they can rightly expect a reward for behaving well, but that they are ultimately able to make sense of the social outcomes of their deliberated actions.

The world in which the truly just person is systematically reviled and thwarted is thus one in which the moral order has been turned on its head. But every agent relies on this order. In the real world this is true even of the truly unjust person—the one with a reputation for justice—idealized by Glaucon and his aristocratic friends. Even tyrants need other people to help them achieve their ends, though their actions gradually reduce the number of people upon whom they can reliably depend (think of Richard III).

The same goes for the just person with a reputation for injustice. This agent's notorious 'reputation' is a stand-in for a world that has become fundamentally recalcitrant to the exercise of virtue. Stoic philosophers and Kant face the problem of a radically recalcitrant world by shutting it out. When moral rationality is reduced in this fashion to conforming the will to an abstract principle—the categorical imperative or Stoic virtue—there is no need for one's actions to be taken up by the world. In this case, despite its brutality the world described by Glaucon—where the just person is 'whipped, stretched on a rack, chained, blinded with fire, and, at the end...impaled' (Plato 1992, 361e, 37)—remains morally intelligible.

As I have been arguing, however, this is not intelligible in *our* moral world. To the contrary, it is radically unintelligible. It expresses a deep alienation from social reality and is contrary to basic Enlightenment goals and therefore also the ideals of the Anthropocene Project. Although at bottom the characters he describes are probably unrealizable in their pure forms, Glaucon's perverse fantasy does reveal the all-too-real possibility of moral chaos. And this is the sort of world we may be headed towards because of our inaction on climate change. We have plenty of experience with these kinds of moral breakdowns. Consider the letter sent by seven Tutsi pastors to their 'Dear Leader, Pastor Elizaphan Ntakirutimana', at the height of the Rwandan genocide:

> How are you! We wish you to be strong in all these problems we are facing. We wish to inform you that we have heard that tomorrow we will be killed with our families. We therefore request you to intervene on our behalf and talk with the Mayor. We believe that, with the help of God, who entrusted you with the leadership of this flock, which is going to be destroyed, your intervention will be highly appreciated, the same way as the Jews were saved by Esther. We give honour to you. (Quoted in Gourevitch 1998, 42)

Because of the way it is combined with a lucid awareness that the world is going brutally to pieces, the studied observance of epistolary niceties here is remarkable. The odd juxtaposition highlights the fundamental unintelligibility of what is happening from the standpoint of the victims. Even as they grasp what is likely about to unfold, the pastors clearly *expect* specific other people—the Mayor, their Leader, God—to intervene on their behalf and stop the *génocidaires*.

Unless we have already succumbed to a numbing moral nihilism, we cannot help but think that the world simply should not unfold this way. Imperfectly good people should be able to rely on the goodness of others in times of crisis. This is what Strawson means in saying that 'we *demand* some degree of goodwill or regard' from specific others, and not just intimates (2008, 7; my emphasis). This is a basic feature of social relations and is therefore something we should be emphasizing in discussions of human security and insecurity.

The demand in question is not aimed primarily at people's actions—still less the consequences of those actions—but at the quality of will behind them. The pastors' letter implicitly expresses a demand for *stability* in the pro-social dispositions of certain authorities. They are not deceived about the threat they face, but they assume—or hope—that those in a position to help will be reliable even in the teeth of massive countervailing situational pressure. Other things equal—that is, assuming the people appealed to were not simply prevented from helping, which might have been the case (except for God, of course)—the failure to do so will be seen by the letter's authors as a *betrayal*. It is a negative judgement about the instability of character of those who could have done more to help had they remained firm.

Think of this in relation to the *Stern Review*, analysed in Section 3.2 above. As I have said, reading this report, one might conclude that although the task of future adaptation will be difficult indeed, and from a purely prudential point of view it would certainly be better to divert 2 per cent of global GDP to tackling the problem now, it is essentially manageable. So we should not be too worried. But what no economic calculus can hope to capture is the colossal weight of pain, sorrow, depression, stress, anxiety, and crushing world-weariness associated with the collapse of livelihoods, the premature deaths of loved ones, forced migration into strange new lands with possibly hostile inhabitants, the sheer brutality of eking out an existence in a withered landscape, the knowledge that one's neighbour may turn on one suddenly, the loss of social and cultural capital, and so on. Although institutional failures are a key causal component of these losses, from the standpoint of the victims it is no less important to say that they are the product of moral failure on the part of other members of their community. That is the real tragedy. To counter it, all of us—but precisely in proportion to the severity of the crises of adaptation we will face—are going to require the sort of moral indemnification provided by firm character.

This brings us to the duty of mitigation, the main point of which is to minimize harms visited on people of the future through extreme climatic degradation. It is tempting to articulate this duty in terms of rights, but I want to suggest that this concept says too little about the failure involved in our allowing a morally chaotic world to emerge. One of the reasons we have been so slow to move away from climate change business as usual is that we are operating with an outdated conceptual framework of rights and asso-ciated responsibilities. Rights discourse is important and good work is being done to show how and why we should think of future people as bearers of rights we are in the process of violating.[12] There is no reason to abandon this approach altogether. However, at least on this issue it appears to have run out of steam and we might wonder why that is. Perhaps the problem is that rights-talk seems peculiarly prone to a certain kind of self-serving mischief.

We are accustomed to the idea that rights need to be balanced but tend not to notice the frequency with which in doing so we place a thumb on the scales.

That is, even if we in the present came to accept the notion that future people have rights, and to assent to the claim that there is, as things currently stand, a conflict between some of our 'rights' and some of theirs, what would this change? In this case we would likely see the rights of future people as quite spectral things, the same way many people now think of 'animal rights'. Unfortunately, the point applies as much to basic rights—to life, health, and subsistence, for example—as non-basic ones. It's a pity that factory-farmed animals are treated badly and then killed, but, given our numbers, our 'right' to feed ourselves adequately—which, so goes the argument, requires factory farms—overrides their 'right' to a life (and death) free of extreme pain. Although it has achieved relatively deep cultural penetration over the last forty years, and despite some improvements in our treatment of non-human animals in that time, appeals to animal rights have not led to substantial, global reform of factory farming. Evidently, in the way we treat both non-human animals and people of the future it is simply too easy to say that in going about our business we are 'acting within our rights'.

If we are going to get anywhere in our defence of the vital interests of future people, we need to try a different approach, or at least to see how we might supplement rights language. Thomas E. Hill Jr (1983) argues that people can be criticized morally even where their actions violate nobody's rights. In the example he provides, someone moves in to the house next door to yours and proceeds to destroy the beautiful, biodiverse garden in the backyard so that he can cover the area in asphalt. Though the man acts 'within his rights', the question you presumably want to ask is, 'what sort of person would do a thing like that?' Similarly, Hursthouse has argued that in talking about abortion it is insufficient to invoke rights, because even if one acts within one's rights (in getting an abortion, say), one might 'do something cruel, or callous, or selfish, light minded, self-righteous, stupid, inconsiderate, disloyal, dishonest' (1991, 227). Hill's argument is applicable well beyond environmental ethics and Hursthouse's well beyond the ethics of abortion. Both philosophers are claiming, more generally, that even where rights claims make some sense they do not always fully capture what we understand as moral failure. In assessments of wrongdoing, pointing to what agents *are* goes deeper than simply pointing to what they *do* (or fail to do). This is not to say that moral criticism of them is always aimed at showing that they lack a particular virtue *tout court*. Instead, we might say that a person has failed, for one reason or another, to deliver a virtuous performance in a situation in which we have rightly come to expect one of her because, in similar situations in the past, she has performed virtuously. In other words, as was the case with the morally chaotic world described above, the failure may have to do with the stability of her character across situations and circumstances.

This point brings us back to the critique of situationism from Chapter 1. Situationism provides real insight into the banality of evil, but, I argued there, it does not establish either the impossibility or the ethical undesirability of stable dispositions of character. The current point is that it is crucial for those with strong duties of mitigation (the global prosperous) to develop the stable disposition of justice. Many of us already engage in desire-constraint out of a sense of justice. Buying fair trade coffee even though the other stuff tastes just as good and is cheaper is just one example, but virtually everyone performs numerous actions like this. The phenomenon also shows up in what we consider to be out of the question. We don't own slaves or buy replacement organs on the open market, and not *only* because there are laws against these practices. The latter are instructive examples because they indicate that there can come a time in our social history when desire-constraint no longer feels like it used to with respect to some object of desire. It is now *easy* to refrain from doing certain things because desire has been tamed or trained by considerations of justice.

So we all know what it *means* to constrain desire on principle. Moreover, this activity is reason responsive: assuming the accuracy of the information we get about the condition of South American coffee growers, we buy fair trade coffee *because* we want to promote the economic flourishing of these communities, especially in an industry notorious for the exploitation of some of its suppliers. But what if the evidence now suggests that much more far-reaching desire-constraint is required if we want to meet the requirements of specifically intergenerational justice? I've already made the argument—in Chapter 2—that the evidence does indeed suggest this, but here I want to note that what is required is the same virtue (or set of virtues) we already know how to use. What we are asking it (or them) to do is remain firm in the face of the challenging new circumstances of justice.

In this sense, we are at a point in our social history analogous to a certain period in the history of abolitionism in the United States. We *could* say that in the middle decades of the nineteenth century the whole American population was divisible into two groups: those staunchly for abolition and those just as staunchly against it. But that is crude social history. Imagine all the people in between—for instance those in so-called 'border states' like Kentucky and Tennessee—who might have *felt* the contrary pulls of easy money and tradition on the one hand and racial justice on the other. At that time, many Americans needed to learn how to constrain a powerful subset of their desires, and they succeeded in doing so (albeit with the help of a war). Among many other things, this can be considered a victory of dispositional stability. Managing the motley forces of *our* psyches will be no less daunting. As we put our backs into the difficult task of mitigation some of us will need new virtues, others will doubtless require coercion, but many more of us will simply need to learn *firmness* in the application to new circumstances of virtues we already possess.

A final consideration is that appeals to virtue and vice can be powerful motivators for agents.[13] Most people genuinely dislike being called vicious. To the extent that she takes the charges seriously, no decent person can remain unruffled while having even a few of Hursthouse's epithets applied to her. Our abject failure (so far) to mitigate indicates that, whatever we claim about our generational rights, people of the future may be justified in calling us callous, reckless, indifferent, and cruel. Indeed, from that perspective our behaviour might, as Gardiner suggests, reveal a 'fundamental incompetency in agency', even though what we are doing feels perfectly normal to us (2012, 249). In subsequent chapters, I will recur sometimes to this sort of perspectival switch, placing future peoples' plight at the centre of the analysis and asking what our characters look like from there. The task is to deepen the critique of the present, moving from a description of the catastrophically transformed world our behaviours will leave behind to the moral assessments people of the future are likely to make of us because of this.

3.6 Conclusion

One might suggest that a longer view than the one taken here is required. Perhaps in a few hundred years we will have reduced our population dramatically and will have acquired some valuable lessons about sustainable practices along the way. Perhaps. But we cannot know that things will turn out that way and even if they did it would not be enough to justify the path we are taking to get there. In his brilliant novel, *Blindness*, Saramago has one of his characters making the key point, a more colourful version of the claim Pachauri makes in summing up AR5's assessment of the future security situation: 'We're going back to being primitive hordes, said the old man with the black eyepatch, with the difference that we are not a few thousand men and women in an immense, unspoiled nature, but thousands of millions in an uprooted, exhausted world (1995, 255–6). We are a world civilization comprising billions of highly interconnected individuals and the danger we face is that this skein of humanity will begin to unravel in ways for which we have not planned and which will therefore challenge profoundly our capacity to coexist peaceably. But I have also argued that facing the danger squarely and calling it what it is—a catastrophe—is no summons to fatalism or pessimism. We are still in a position to change things and avert the worst effects of climate change.

Our failure to mitigate will make adaptation for people of the future that much more difficult. If they fail to do so in morally salutary ways, significant blame can be laid at our feet. For these reasons, in what follows I will analyse our failures—those of the global prosperous—to take seriously our duties of mitigation as indications of underlying character flaws, both moral (Chapter 4) and epistemic (Chapter 5). This focus no doubt runs the risk of its own kind of distortion. But if we think of the relation between present and future generations as one of oppressor to oppressed, the distortion can work both ways. As Claudia Card has argued, while victims of evildoers tend to exaggerate the 'reprehensibility of the motives of perpetrators', the latter tend to underestimate the magnitude of the harms they cause (2005, 485). Given that the material benefits of our current practices—burning all that fossil fuel—are weighted so heavily to the interests of people of the present, and that this is all currently rationalized in transparently self-serving ways, it is worthwhile to risk moving to the other perspective, that of the victims of climate change. In surveying the shattered world we have given them, how will they answer the question, 'what sort of people would do a thing like that?'

4

Justice

4.1 Introduction

Michael Beard, the protagonist of Ian McEwan's novel *Solar*, is a thoroughly unlikable fellow. He's also a climate scientist and entrepreneur. With this character McEwan seems to be telling us that we are essentially hypocritical or insincere about climate change because although many of us are uneasy about the problem and want something done about it, we are at the same time slaves to our appetites. After contemplating the possibilities of getting involved in a potentially lucrative geo-engineering scheme (dumping iron-filings in the ocean) that would allow him to make substantial money trading carbon credits, this is how McEwan has Beard settling down late one evening:

> [T]here was much to consider. Between a radicchio lettuce and a jar of Melissa's homemade jam, in a white bowl covered with silver foil, were the remains of the chicken stew. In the freezer compartment was a half-litre of dark chocolate ice-cream. It could thaw while he got started. He took a spoon from a drawer (it would do for both courses) and sat down to his meal, feeling, as he peeled the foil away, already restored. (McEwan 2010, 188–9)

Beard's desire to save the planet exists in him cheek-by-jowl with an appetite that feels enormous to him and looks that way to others too. His greed is gluttonous, and insofar as he is meant to be a stand-in for the rest of us, so is ours. He's a hypocrite because his environmentalism is a product of his greed: he uses the climate crisis as a way of getting rich. McEwan is right to juxtapose these two forces in us—endorsement of a just cause and greed—but is incorrect to suppose that hypocrisy or insincerity is our main problem. It's our weakness we should be most worried about and this is importantly different from the disorder McEwan identifies.

Both insincere agents and morally weak agents can be said to endorse a moral principle, P. However, the endorsement of the insincere agent is merely hypothetical because it is dependent on the extent to which she believes that serving, or seeming to serve, P advances an independent desire or interest. Commitment to P disappears insofar as the agent ceases to believe that it will serve the desire or interest. Were Beard to come to believe that climate change presents no opportunities for pecuniary gain he would quit seeing it as a crisis. The morally weak, by contrast, do not endorse P hypothetically. It is embraced categorically, but the agent is able to 'see' that the demands it makes on her are too difficult to face squarely. So she pushes them away and acts in ways that run contrary to them. The task of Chapter 2 was to identify the foundational principles of justice for the Anthropocene. An agent must take such principles as reasons to qualify as intergenerationally just, but the principles will get her nowhere if she is morally weak or lacks integrity. Strength of soul and just convictions are necessary and jointly sufficient conditions for being a just person, so in this chapter we will focus on strength of soul in the context of the climate crisis.

I begin by highlighting the importance of the vice of greed. If we are to characterize ordinary agents as unjust in the right way, the key is to cease thinking of greed as a kind of feeling and to adopt instead a more objective view of it. If we do this, it becomes evident that we are greedy just to the extent that we are, in fact, taking more than our fair share of a scarce resource. This often feels like normal, prudential behaviour, but that is misleading. My next task is to make the case that our inaction on climate change is largely a matter of moral weakness. It should come as no surprise that the way we consume is at the heart of the analysis. But I add to this familiar claim the idea that consumption is a conspicuous enabler of attention-diversion, a description which fits the Aristotelian notion of moral weakness. Further, since so many of us think that we can solve the climate crisis by bringing more efficiency to our exchanges, I also argue at length against this approach. The problem with it is that it mistakenly supposes we can move towards justice without meaningful desire-constraint. But we require psychic allies in the task of desire-constraint, which is why I move, finally, to a consideration of what the Greeks identified as the 'middle part' of the soul. I provide an interpretation of the moral psychology of shame and honour, emphasizing both the constructive aspects of these emotions and the way they can be deployed in the service of Enlightenment values.

4.2 Justice and Greed

Climate ethicists have examined the problem of intergenerational justice almost entirely in terms of distributive, procedural, and retributive justice.[1] I don't want to gainsay the importance of this work, but it does overlook an alternative approach to justice that looks to issues of character and disposition rather than outcomes or procedures. Although the former approach is more centrally tied to premodern conceptions of justice, it remains relevant. For example, in a recent study of justice, Sandel argues that an illuminating way to understand practices like price-gouging in times of crisis is that the practice is, fundamentally, greedy and is thus contrary to 'civic virtue' (Sandel 2009, 6–7). At times like this we require a sense of solidarity among citizens. And here the question Hill's agent asked about his asphalt-loving neighbour (examined in Chapter 3) seems apposite: even if price gouging were not illegal in times of crisis (so the gouger was acting 'within his rights' in hiking his prices), we think he has nevertheless failed morally. And the best way to understand this reaction, and the outrage that often accompanies it, is that it is a negative assessment of the person's character.

But what exactly are we saying about his character? First, that some of his dispositions are best characterized as vicious. In this agent at least, we are supposing that greed is a more or less stable character trait that, because it is immoral (that is, harm-causing), is also a vice in him. Beyond this, however, there are interpretive options. Think of this in terms of Aristotle's discussion of moral types. For Aristotle, non-virtuous agents come in three forms: the strong, the weak, and the vicious. The strong have to fight contrary-to-virtue impulses and desires in order to do the right thing, but always manage to do so. The weak undergo the same internal struggle, but always succumb to the wayward impulse or desire. And the vicious have bad dispositions, which they, qua vicious, invariably follow.[2] If Aristotle is correct then we should criticize strong, weak, and vicious agents differently. We may not be inclined to criticize the strong at all, though Aristotle certainly did. But clearly the weak fail in the job of proper self-constitution while the vicious fail to recognize the authority of external moral standards. In the former case, the main problem is likely to be volitional while in the latter it is likely moral-epistemic. And note that we need to know quite a bit about an agent to decide which of these criticisms applies to her in cases of moral failure.

For example, from the scant information he gives us we don't know exactly how to round-out our picture of Sandel's gouger. Perhaps his greedy desire overwhelms him and obstructs from his view what he otherwise knows he should do. This is volitional failure. Or maybe he really believes he should take advantage of people's distress in a time of crisis by over-charging them for basic necessities. He just cannot 'see' what is wrong in this, so his failure is moral-epistemic. To complicate matters even further, our knowledge here will probably always be incomplete. Because we had always thought of him as a conscientious member of our community, the gouger's actions may be contrary to our expectations of him, which may indicate that he is acting weakly. But maybe he does not experience any inner turmoil at all. He might see himself as a particularly fastidious upholder of the law of supply and demand, in which case he believes it necessary on principle to raise his prices in response to an uptick in demand for his product. It's just that prior to the crisis he (and we) hadn't imagined the principle would ever be extended in this way. At some point our interpretive spade hits bedrock and we have to go with our best explanation of a person's actions.

Aristotle's account is insightful, but we need to modify it in two ways. First, we should reject his claim that the strong person is a moral failure. It might be better if we could all be perfectly virtuous and that might even be an ideal for some of us, but for most of us strength is good enough. This matters because if we set our sights too high, we might inevitably fail and allow demoralization to set in. This is a particularly pressing worry with a moral problem as forbidding as climate change. Second, weakness can result in action that is every bit as immoral as the action consequent on vice. For this reason, we should characterize moral weakness as a form of vice, and reserve the term 'wickedness' for what Aristotle calls vice. If we ask whether our failure to do very much so far about climate change is due to wickedness or moral weakness, I think the inescapable answer is both. Some of us are best described as wicked, others merely morally weak (it is also possible for a single agent to display both faults). The distinction is important because how we assess the failure affects the shape of our reactive attitudes towards the wrongdoer. Generally speaking, since she has internalized the correct norm, the weak agent is more morally pliable than the wicked one. Other things equal, this means she is relatively amenable to prospective moral correction as well as processes of moral repair for past wrongdoing.

By contrast, although some of them can surely be brought around, the problem with the truly wicked is that they are likely to dismiss moral criticism of them as browbeating. For example, Bill McKibben (2012) has argued that in the age of climate change, fossil-fuel company executives, and the politicians who abet them, are the clear enemies of future humanity. It looks as though he is saying that in their heedless pursuit of profit such agents are wicked, perhaps incorrigibly so. This seems like just the right description of Charles and David Koch, for instance. Koch Industries, a fossil-fuel consortium, has contributed roughly $67 million to climate change denial campaigns over the years. Though it is more controversial to do so, one might also point to Bjørn Lomborg here, since he seems not at all disturbed by the fact that his 'environmental skepticism' has been so influential in the ideological struggle against decisive mitigation measures. To my knowledge, nobody targeted by this sort of critique has been moved by it (except possibly in the direction of modifying PR strategies), which suggests that the thesis about the incorrigibility of the wicked is sound.

Justice as a virtue is the disposition to accord each member of the moral community his fair share of socially distributed benefits and burdens. But it is more than this. It is also identity-forming. To be just is to see oneself as part of a larger social whole and to care about the integrity or health of that whole. To be intergenerationally just is thus to see oneself as part of a whole that includes future people. Character traits impart a measure of inflexibility to one's identity (Jamieson 2007). They require one to draw moral 'lines in the sand': to consider unjust actions, for instance, unthinkable. Moreover, as O'Neill argues, even with the best institutions in place, the virtue of justice is required because no 'institutions are...knave-proof' (O'Neill 1996, 187). Among the vices, the most potent threat to justice is greed, since this has centrally to do with taking more than one's share of a good. But climate ethicists are divided about whether or not our inaction on climate change is the product of greed. Gardiner (2012) and Kawall (2012) think that invoking such vices is entirely appropriate. Dale Jamieson appears to disagree: 'We should remind ourselves that while a great deal of environmentally destructive human behaviour can rightly be denounced as greedy or vicious, much is humdrum and ordinary...Many of our environmental problems have the structure of collective action problems (2012, 231). The implication here is that we can explain our moral failure on climate change *either* as a collective action problem *or* as the upshot of greed, but not both. Vogel concurs with Jamieson, claiming that the problem with agents in the throes of a collective action problem is not their greed but their 'isolation' from one another, isolation imposed and maintained by the institutions of the free market economy (2012, 308). This is a misguided approach to collective action problems. By contrast, Parfit gets the matter right:

> It is not enough to ask, 'Will my act harm other people?'...I should ask, 'Will my act be one of a set of acts that will *together* harm other people?' The answer may be Yes. And the harm to others may be great. If this is so I may be acting *very* wrongly...We must accept this view if our concern for others is to yield solutions to most of the many Prisoner's Dilemmas that we face: most of the many cases where, if each of us rather than none of us does what will be better for himself—or better for his family or for those he loves—this will be worse, and often *much* worse, for everyone. (1984, 181)

Because it is possible to do a great deal of moral wrong in the context of collective action problems we require recourse to the strong moral language of the vices in order to diagnose how agents go astray in them. What is the best way to think about this?

Agents in collective action problems are standardly characterized as rational choosers, and rationality is honorific. Thus even though these situations usually produce suboptimal outcomes, it has always been problematic to know how (or if) to reproach an agent caught up in one of them so long as she calculates her advantage accurately. Moreover, she will likely deny that her decisions spring from greed—because it does not feel that way to her—and, lacking a more authoritative heuristic, we tend to take her at her word. More often, agents like this will experience themselves as making self-protective decisions. But why should we place so much weight on the phenomenology of vice? To be sure, greed sometimes has a distinct and very powerful feel to it, for instance when it takes the form of gluttonous desire (think again of McEwan's Beard). It may be that philosophers like Jamieson and Vogel are implicitly taking gluttony as the paradigm of greed. But since the 'soccer mom', for example, fails to exhibit the vice in this form—she might even engage diligently in such 'green' conscience-salving activities as recycling her household wastes—it must be necessary to say something else entirely about her: for example that in consuming the way she does she is only protecting what she and her family need in order to live a good life.

But can we not say that she might *also* be greedy or that one form that greed can take is the desire for enhanced economic security? Think of the issue in terms of the paradigm example of the prisoners' dilemma (cops and robbers). It's highly artificial, but there's much to learn from it. Both prisoners in this game are driven by two psychological forces: the desire for gain and the fear of loss. From either the agents' standpoint or from that of an impartial observer it is not easy to separate these forces in order to determine which is doing the motivational heavy-lifting. And where we are chiefly guided by the desire for gain, we might rationalize our choices as being entirely defensive. Note what 'gain' means here. Each prisoner is being offered no jail time at all if she defects while the other does not. But here's the important part. As someone who has by hypothesis committed a crime (the lesser one), for which the police could convict her, this is *more* than she deserves, even though she likely tells herself that in seeking the sweet deal she is simply securing what she takes to be her morally unproblematic *ex ante* position.

In Chapter 3 I argued that we all seek a certain baseline of security in our lives, that we expect other people and our institutions to respond reliably and intelligibly to the things we do. But the search for security has a less laudable face too. Gabrielle Taylor has argued that the key feature of the avaricious is her 'desire for guaranteed security of position' (2008, 37). While deliberating with others one can *feel* as though one is only seeking security of position, when one may in fact be in the process of acquiring much more than one needs. In other words, greed and anxiety/self-protection are often two faces of the same psychological coin. So with scarce resources, where acquisition at the margin is a zero-sum game, anxiety and self-protection may be working against the claims of justice. It follows, contra Jamieson and Vogel, that greed might be operative in collective action problems. This way of putting the point fits the story Hirschman tells about the way in which the eighteenth century gradually put avarice in the 'position of a privileged passion', one that came to be associated explicitly with 'interest'. This allowed the process of relatively unfettered accumulation to assume a socially benign aspect through its connection with the unremarkable virtue of prudence (1977, 41).

Compare these ideas with what some feminists have said about the moral psychology of members of dominant (as well as dominated) groups. In a superb study of the 'burdened virtues' that characterize both oppressors and oppressed in a patriarchal society, Lisa Tessman argues, 'Given the pervasive injustice of oppression and given the high level of participation in maintaining structures of oppression and the difficulty of unlearning traits associated with domination even for those who become critical, I see unjust and other vicious people as fairly ordinary' (2005, 56). This is important for two reasons. First, as Tessman points out, it corrects the mistaken view of traditional virtue ethics (to say nothing of consequentialist, contractarian, or deontological ethics), according to which most of us are virtuous or at least not 'steeped in vice', as Hursthouse puts it (Tessman 2005, 55). If Tessman is correct the problem with these views is their abstraction from the real workings of power. Because oppressors need to rationalize their actions, the fact of domination, especially where it has been operative for long periods of time, typically distorts their sense of virtue.

Second, because of this the feminist position fits the climate issue perfectly. Think, in this connection, of Gardiner's notion of the 'pure intergenerational problem' (PIP), which defines the stance of many current individuals, firms, and nations, especially in the developed world.[3] According to the PIP, members of the current generation engage in significant 'intergenerational buck-passing'. That is, because our interests are almost entirely generation-relative, we engage in actions—especially those associated with fossil fuel–based consumption—whose benefits accrue largely to us but many of whose costs will be borne by members of the future. We fail to invest in what Gardiner calls 'back-loaded goods', those efforts—like sharp reductions in GHG emissions—whose benefits accrue to future generations but whose costs are borne mostly by present generations. For Gardiner, the PIP thus constitutes a 'tyranny of the contemporary' over the future.[4] Is it any surprise that we do not yet see the PIP as supported by vices like greed nor indeed that in conducting the business as usual that *is* the PIP we typically think of ourselves as virtuously prudent—even compassionate and generous—people?

Against the phenomenological conception of vice, we should insist on a more statistical and comparative notion. For example, since there is a tight connection between GHG emissions and levels of consumption, we should look to data about per-capita GHG emissions in order to fully understand the global topography of justice and greed. Here we find such huge discrepancies between, say, Australia and Morocco—in 2009, Australia's per-capita GHG emissions were 19.64 tonnes, those of Morocco 1.17 tonnes (Rogers and Evans 2011)—that we are prima facie warranted in talking about the average

Australian citizen as lacking the virtue of justice. Again, from this perspective it is irrelevant that this or that Australian does not feel greedy in the choices she makes, since she is as a matter of fact using more of a scarce resource than justice demands. The same point can be made about the present generation as a whole and future generations. Since we are using far more than our fair share of the planet's capacity for carbon storage, we are acting greedily vis-à-vis people of the future. The proof has nothing to do with how our behaviour feels to us and everything to do with our steadfast refusal to work towards decarbonizing the global economy.

In the Anthropocene, what looks like ordinary behaviour has become deeply problematic. Melissa Lane has argued that the Greeks were more attuned than we are to the problems of *pleonexia*, the overweening desire for gain. A good deal of Greek philosophy, as well as Enlightenment appropriations of it, was focused on the ways in which this socially and politically corrosive desire could be constrained. But the age of fossil fuels introduces a new challenge because the energy these fuels unleash removes 'the final constraint on *pleonexia*' (Lane 2012, 40). One need only read Daniel Yergin's magisterial two-volume history of oil to appreciate the truth in this (2008, 2012).[5] The age of oil *is* the age of frenzied consumption. In the eighteenth century, when thinkers, religious authorities, and political leaders were all struggling with ways of curbing anti-social passions like glory, it seemed like a good idea to allow greed wider scope, but now it too is proving to be a danger to the social and natural fabric. My point about the way greed can work in the service of seemingly positive forces like the quest for economic security is that we need to begin to see pathology in cases of what Lane, echoing Tessman, refers to as 'apparent success'—soccer moms and corporate CEOs—no less than with more familiar cases of social pathology (2012, 109).

Is this too revisionary? I don't think so. What I am saying is that it is not simply up to agents themselves to define the conditions of application of the vices (or virtues). Greed is constitutively tied to justice, and the demands of justice are objectively specific to a time and place. They have to do with the relative scarcity of vital resources faced by a particular group of people at a particular time. In times of moral crisis—when a new appreciation of scarcities forces a group to distribute benefits and burdens in a way not practised or perhaps even contemplated before—many members of the group will doubtless feel as though they are being mistreated. If they find themselves relatively deprived in the new dispensation, they will protest that there was nothing amiss with their pre-crisis appetites and will therefore claim that they are not being treated fairly. But assuming the new dispensation conforms to the demands of justice the claim is simply not true given the crisis. Tessman's point is that in situations of oppression, 'disordered appetites' in the oppressors are the norm, for precisely which reason they are not perceived to be the source of outlandish demands by oppressors themselves. If, with Gabriele Taylor, we can get used to seeing avarice as closer to the desire for security than it is to gluttony, we will have made real headway here. But to say that our greed is a vice is explicitly not to say that we are wickedly greedy. Instead, greed is a manifestation of moral weakness, a phenomenon whose role in helping to perpetuate the climate crisis we need to examine next.

4.3 Moral Weakness

We are morally weak when we choose (and then presumably act) in ways that run contrary to our best judgement about what we should be doing. The smoker who wants to quit but lights up another one anyway, the person who knows he should be going to the gym but decides instead to spend the afternoon watching television, the sometime political activist who realizes she should be out today canvassing for her local candidate but who can't resist taking in that movie she's been dying to see. And so on. There are two fault lines in discussions of weakness of will. The first one dominates the history of philosophy since Socrates and has to do with the possibility of weakness of will. The position of Socrates himself describes one side of this division because he thought that, since virtue is knowledge and the will always follows the apprehension of the good, weakness of will is not possible. The other side takes its inspiration from St Paul, arguing that weakness of will is not only possible but is unavoidable for postlapsarian humanity.[6]

Both the Socratic and the Pauline views are too extreme, though arguing for this would obviously take me too far afield. So in what follows I'm simply going to assume that weakness of will is possible but also avoidable in most cases. The more interesting fault line assumes in this fashion the plausibility of some middle position between Socrates and St Paul and has to do with the mechanics of moral weakness. On the one hand, the examples I have just sketched might be described as cases of 'strict' or 'open-eyed' weakness. On this interpretation of them, the agent acts contrary to her considered judgement in full awareness of what she is doing. There are, on the other hand, those who reject this view, arguing that if we can act weakly it is only because we manage somehow to withdraw our attention from our considered judgement in the course of acting against it.[7] When this happens we are usually beguiled by desire: we think about that sweet smoke, those blissful hours on the couch, the handsome face of the actor in the movie. In what follows I will adopt the attention-withdrawal model of weakness—thus rejecting the notion of strict or open-eyed weakness—to help illuminate the mechanics of moral weakness as applied to the failure of the global prosperous to act meaningfully to meet the challenges of climate change.

Socrates is surely on to something in insisting that it is difficult to understand how we could intentionally act contrary to our best judgements. It seems as though we necessarily act under the guise of the good, that is, on the assumption that we are doing the best we can in the circumstances. If this is right, and we also want to save the appearances by recognizing that we are sometimes weak, the attention-withdrawal model of weakness of will looks attractive. This is where Aristotle becomes important. Consider his critical appropriation of the Socratic position on weakness. Again, Socrates argues that to affirm the possibility of strict or open-eyed moral weakness is to assume, absurdly, that knowledge (of the good in this case) can be 'dragged about like a slave'.[8] Aristotle wants to account for our intuition that we are sometimes weak without allowing knowledge to suffer this sort of indignity. According to Francis Sparshott, this squares the Socratic circle because when we are weak, 'something prevents us from effectively noticing the aspects of the situation that should bring it within the scope of our moral convictions. So Socrates was, in a way, right after all. Knowledge is not dragged about, it is simply bypassed' (1994, 249). This, I submit, gets things exactly right. Applied to our case, the point is that our desire to consume more than we need allows us to 'bypass' the knowledge that such consumption violates our moral convictions. Consumption in capitalist societies is essentially a means of attention-diversion. In Chapter 3 I quoted Jameson's helpful view of reification: we reify objects of consumption insofar as we efface from them all past traces of their production as well as the effects of their production on the future. The largely short-term pleasure they give us is tied to their ability to obscure issues of justice with which their production and consumption are bound up. This has a great deal to do with how such objects are advertised to us, of course. We believe that the object's meaning is exhausted by what its pushers say about it, which is invariably focused on the bogus link between our consumption of the product and our happiness. The problem with this structure is that it is impeding our capacity to live with integrity. How should we understand this?

One of the most important contributions of Greek ethics to our understanding of justice is that the Greeks, especially Plato, saw justice as a matter of proper self-constitution, at both the agential and collective levels, that is, they saw it as a matter of integrity.[9] According to Woodruff, this means that justice cannot be 'merely mind-deep' but must reach into the soul's sub-layers and draw together the forces operating there.[10] Because of this, self-mastery is the guiding moral-psychological ideal. Commenting on how ridiculous the idea of self-control seems, since in this case the same person is both controller and controlled, Socrates argues that the expression nevertheless has significant moral import and explanatory power:

> [T]he expression is apparently trying to indicate that, in the soul of that very person, there is a better part and a worse one and that, whenever the naturally better part is in control of the worse, this is expressed by saying that the person is

self-controlled or master of himself. But when, on the other hand, the smaller and better part is overpowered by the larger, because of bad upbringing or bad company, this is called being self-defeated or licentious and is a reproach. (Plato 1992, 431a)[11]

If this picture is applicable to us we need to name the forces blocking our self-mastery. At the risk of oversimplification, let's state the conflict in the starkest possible terms, adding nuance as we go along. On the one hand we endorse the cosmopolitan ideal extended to include future people, and because of this we are, as Elizabeth Cripps puts it, a 'should-be' collective comprising the present and the future.[12] On the other hand, as Adrian Parr argues, 'the law of value' operative today is that of the individual consumer whose choices are sovereign (2012 , 64). This 'law' not only licenses but demands unfettered consumption, while the cosmopolitan principle demands significant constraints on it. When resources are scarce and have multiple claimants, those who already have more than their fair share cannot be just without self-constraint.[13] That is the chief insight of the Socratic approach to justice.

As it stands, this is not a fair fight, precisely because of what Socrates said: the larger force in us, the desire for more than we need, can easily overwhelm the smaller part. It's worth emphasizing that, as I have been arguing, the main source of our attachment to people far-flung in future time, to the extent that we appreciate it at all, is cognitive. We recognize, or should, that we have a duty to avoid harming people of the future as a logical extension of our duty to avoid harming anyone whose vital interests are affected by our actions. Because the appropriate actions to take in light of this duty involve consuming less than we currently do, it is understandably difficult to summon the motivation to do the right thing, especially since our culture bombards us with exactly the opposite message. If this is correct, it means that there is a good deal of prima facie plausibility to the idea that we have largely botched the job of self-constitution. So the question then becomes how to repair the damage. Our answer will necessarily be somewhat schematic, but even so it indicates a way forward.

In Section 4.2 above I argued that ordinary desires to get ahead and prosper can be greedy even if they are not experienced this way by greedy agents themselves. If this is correct it means that such agents are taking more than their fair share of a scarce resource. This is a bad thing from a purely impersonalist point of view, but it also runs contrary to moral convictions the global prosperous already have. To support this claim, imagine how you, a member of the global prosperous, would respond to Henry Shue's example of a person planting land mines on a busy pathway and setting them to explode sometime in the far future. You can doubt that this is a good analogy to what we are doing with climate change, but in light of all the evidence we now have about the future effects of our action and inaction that gambit is unconvincing.

As Shue says of his bomber, 'it is *obviously* totally wrong' for a person to do this (Quoted in Cripps 2013, 75; my emphasis). The qualifier 'obviously' is doing a great deal of philosophical work here: it is simply not sane to deny the wrongness in the example. However, by extension, we must grasp the wrongness of climate change and the greed that enables it.

It might be objected that most of the global prosperous have not subjected their behaviours to this sort of examination and that they do not therefore have a moral conviction concerning the wrongness of climate change. But the inference here is suspect and this is precisely my point in bringing up Shue's analogy. *Why* have such agents not engaged in the sort of simple thought experiment that would show them how obviously wrong it is to damage the future the way they are? After all, as we have seen in the previous chapter, by hypothesis we are agents who generally accept the claim that sometimes it is necessary to constrain desire in the interests of justice. The failure I'm describing therefore has to do with an unwillingness to see inferential connections both among our existing moral convictions, and between the latter and the clear evidence about the negative effects of our actions and inactions. All the relevant conceptual and evidential data are sitting there in the mind, waiting to be pieced together. But we withdraw our attention before the connections have a chance to crystallize, and this is what allows us to choose and act contrary to our convictions (i.e., weakly). So the key is to put our convictions back in the seat of psychic power where they belong. How can we do this?

The key to re-establishing integrity is to bring back Jameson's traces precisely in the act of consumption, or as we contemplate buying something. This is a two-step process. First, we need to make imaginative connections between *this* act of energy-consumption and *that* dire climatic outcome. Next, the self-evidently awful characteristics of these outcomes might awaken our appreciation of the claims of justice arising from those having to live with them. Paolo Bacigalupi's dystopian biopunk novel *The Windup Girl* shows us how the two steps might work. The novel is set in the Thai Kingdom in the time of the great Contraction (the near future). The Expansion (our age) has failed due to peak oil, the depredations of agribusiness and the effects of climate change. The City of Divine Beings is surrounded by an ocean steadily on the rise. Water must be pumped out continuously. The political intrigue swirls around two governmental agencies, one of which (Trade) seeks to expand trade, while the other (Environment) wants to curtail it. The struggle to obtain energy—to get through the week, the day, or just the next task—is the central focus of virtually everything the characters do. People eke out a living using whatever energy sources they can find: the last bits of coal, dung, Megodonts (elephant-like creatures), the slave labour of other humans, algae. But life is difficult and good energy is hoarded by the powerful.

Here is how one character is described as she enters a high official's administrative building:

> The computers down here all have large screens. Some of them are models that haven't existed in fifty years and burn more energy than five new ones, but they do their work and in return are meticulously maintained. Still, the amount of power burning through them makes Kanya weak in the knees. *She can almost see the ocean rising in response.* It's a horrifying thing to stand beside. (Bacigalupi 2009, 215; my emphasis)

Like Kanya we need to learn how to make spontaneous, imaginative associations between superficially innocent consumption choices and burning rainforests, rising seas, the destruction wrought by hurricanes, advancing deserts, and so on. I think that if we can accomplish this first step, the next one—connecting the outcomes to claims of justice—should be relatively easy. Kanya's horror at what she sees indicates that she herself is already taking this step. To come back to Shue, completing the two-step process would be one way to highlight the *obvious* nature of the wrongs we do in consuming more energy than we need. Anxiety about the existentially precarious world she inhabits is what allows Kanya to construct the link between the rising ocean and the hum of these energy-sucking machines. There is no reason we cannot spur our imaginations in the same way. In Chapter 5 I cite evidence that our anxiety about climate change is rising sharply of late. Kanya-like imaginative feats of de-reification would be a laudable way to sublimate such anxiety. This is why we need much more climate disaster work from those whose job it is to stimulate the imagination: visual artists, filmmakers, novelists, and songwriters. If my analysis is correct, such work can help bring some unity to our psyches by having our desires face their competitors—our moral convictions—directly. The imagination, thus deployed, can be a powerful ally of integrity and thus of the effort to move from moral weakness to moral strength.

Still, it might be suggested that I'm being too hard on the global prosperous since many of them do understand at least most of what I've been saying here. And they have, largely in response to such considerations, changed the way they consume. They buy more efficient household appliances, drive a car that uses biofuel, and generally try to live in a more energy-efficient manner. This is a powerful lure; one we should resist.

4.4 The Lure of Efficiency

Recently, John Broome has argued that in tackling climate change we should think less about problems of inequity and more about how to increase efficiency by eliminating the externality of GHG emissions. On the Paretian conception of efficiency with which Broome is working, we move from a less to a more efficient distribution of resources when someone is made better off and nobody worse off by the redistribution. Reducing GHG emissions would be efficient in this sense if, through reductions, it were possible to improve the lot of those who suffer from emissions—the poor and people of the future—while not making emitters—the rich and people of the present—any worse off. According to Broome, eliminating the externality of GHG emissions (presumably by internalizing their costs so that the latter are reflected in the price of fossil fuels) increases efficiency, but only if it can be achieved 'without sacrifice' on the part the global prosperous. If this group had to make sacrifices it would be made worse off by the redistribution, making the result Pareto-inferior.

So Broome argues that, because the current rich have displayed no appetite for sacrifice, negotiations should be focused on efficiency without sacrifice. This would involve compensating the current rich for their GHG reductions through a transfer of resources from the 'receivers' of emissions to the emitters themselves. The way this works vis-à-vis the future is that we would be justified in investing less than we otherwise would in natural and artificial resources. This means we can use up more of these resources for ourselves. The result is that we are 'fully compensated' for our emissions reductions, while people of the future are made better off because we have not added dangerously to the atmospheric stock of GHGs.

Here is Broome's conclusion:

> The case I have put for making efficiency without sacrifice an available option is pragmatic. It is not entirely cynical. It is founded on economic rather than moral analysis, but it has the moral purpose of moving the political process forward...
> I hope the availability of the option will get the political process moving, and a better option more like efficiency with sacrifice will be the eventual result. (2012, 48)

It is probably not an overstatement on Broome's part to point out that many people see climate change as a very pressing worry, and know that they should do something about it, but also believe that effective action should require no substantial sacrifice or pain on their part. The idea of efficiency without sacrifice is that we can make meaningful progress towards eliminating the externality of GHG emissions with no downward alteration of our current lifestyles. This is often put in terms of the goal of 'decoupling': the claim that we can break the connection between economic prosperity and emissions, achieving reductions in the latter with no diminution of the former. Although Broome does not invoke the concept, it is clearly implied in the ideal of efficiency without sacrifice. After all, (a) what we are refusing to sacrifice is our standard of living, and this is, as things currently stand, inextricably bound up with steady economic growth; and (b) the contemplated efficiency measures are explicitly aimed at a reduction in our energy consumption. This *is* decoupling. There are two points to make about it and about Broome's larger aim.

First, though in the 'perfect system of transfers' economists fantasize about how it is always technically possible to achieve full Pareto efficiency, it is not clear that efficiency without sacrifice, at least in the form of decoupling, is an achievable goal as the global economic system is currently structured. The case of Sweden is revealing. In 2009, the Swedish government boasted that it had increased its economic output by 50 per cent since 1990 and reduced its carbon emissions by more than 9 per cent in the same period. But a 2012 Swedish EPA report noted that the reductions did not account for 'embedded' emissions. In fact, although domestic emissions fell by 13 per cent between 2000 and 2008, total Swedish emissions, including those embedded in imported goods, rose by 30 per cent. A similar story can be told about Britain and Germany, among other countries (Wijkman and Rockström 2012, 119–21). The problem is that decoupling will not work unless there is no race to the bottom of the carbon-fuelled economy. But since this downward race is still fully operative in the global economy, economic growth and GHG emissions remain strongly linked. Moreover, looking back at the IEA numbers cited in Chapter 3 about future energy use, as things currently stand, by 2035 76 per cent of our energy mix will still be composed of fossil fuels. If the global economy has a fossil-fuel bottom this capacious, and for this long, demand will necessarily have found it.

The peculiar problem with decoupling, for my purposes, is that it invites complacency about climate change by taking the issue of economic growth entirely off the table. In a persuasive analysis of the issue, Danish researcher, Jørgan S. Nørgård has dubbed decoupling 'largely a statistical delusion'. This is aimed at boosters of decoupling who have pointed out that, historically, energy consumption and economic activity do not move in lockstep and that it is therefore possible to break the link between them altogether:

> Every economic activity requires some energy consumption and all energy consumption is rooted in some economic activity, be it on the consumer or producer side. The observation that the two parameters can grow at different rates, which over history is quite normal, does not imply any decoupling... Unfortunately, the notion of a decoupling has served as peacemaker between environmentalists and growth-oriented politicians by conveniently exempting economic growth of any responsibility for environmental problems. (Quoted in Owen 2011, 26)

Looking at the Swedish emissions data examined above, Wijkman and Rockström conclude that we require 'nothing less than a revolution, both in attitudes and in social and economic organization' in order to combat the problem effectively (Wijkman and Rockström 2012, 175). Countries like China and India will continue to produce goods for richer countries, and to produce these goods by burning fossil fuels, so long as that is economically the best option for them. The only way to render it less than the best option is through 'substantial transfers of financial resources from developed countries to developing countries' to assist the latter in developing in green ways (Wijkman and Rockström 2012, 181). These authors thus propose something like Rader-macher's notion of a 'Global Marshall Plan' to help bring about the necessary changes (including, presumably, addressing the issue of lost employment in countries currently producing our consumer goods for us should we succeed in reducing our consumption of those goods). Needless to say, if such revolu-tionary changes are required there will need to be a great deal of *sacrifice* on the part of the current rich. The upshot is that Broome is mistaken in believing that, as things currently stand in the real world economy, we can eliminate the externality of GHG emissions without sacrifice. Making it an 'available option' in our negotiations seems at best pointless and at worst self-deceptive.

This analysis should not be confused with a more moralized condemnation of consumerism. I am not claiming that there is something inherently morally wrong or sinful about consumption, even consumption of 'luxury' goods.[14] Nor, relatedly, am I claiming that decoupling is intrinsically undesirable. To the contrary, if decoupling were possible we should probably pursue it. Afflu-ent countries are now for the most part in a position to move substantially away from fossil fuel–based industrial production. Less affluent countries are, again for the most part, in no position to do so. But if the affluent insist on consuming more than they themselves can produce through green tech-nology, with the difference coming from the fossil fuel–based industry of the developing world, the climate crisis will worsen for the reasons I have been pointing out. The current reality is that we will not lower net global emissions sufficiently unless we *also* reduce luxury consumption dramatically. Decoup-ling is a dangerous fantasy, indeed probably the most dangerous one under which we currently labour with respect to climate change (I reconsider it qua self-deceptive fantasy in Chapter 5).[15]

But suppose I'm wrong about all of this. The second point to make is that even if I am, and we can decouple in time to avert climate catastrophe, we should nevertheless be sceptical about Broome's belief or hope that a politics of efficiency without sacrifice might be a *step towards* a politics of efficiency with sacrifice. Broome himself notes that the main problem with the former is that it is unjust to the current poor and people of the future (2012, 46). There is a substantial motivational gap between the refusal to make sacrifices and the willingness to do so that is simply brushed aside in this analysis. The eco-nomic standpoint as such cannot fully appreciate the problem here. When we say that we will reduce our emissions but by way of compensation also, say, increase our share of the world's fishery resource or decrease the amount we spend on levees and biodiversity protection, we are seeing future people essentially as members of an out-group, as *competitors* with us for scarce resources. By contrast, to the extent that we are willing to make sacrifices to better the lot of people of the future, we are, as Heilbroner puts it, forming 'a collective bond of identity with them' (1981, 138). We will never appreciate the moral value of making sacrifices on behalf of people of the future from within the competitive framework. What we require is a cognitive and imaginative shift, a revolution in our sympathies to accompany the broader social and economic revolution for which some in our culture are calling (a theme to which I return in Chapter 6).

Efficiency is a sop to excessive consumption, a point which brings us back to the problem of moral weakness. Instead of constraining wayward desires or confronting their consequences directly, efficiency-based decisions would have us simply rechannel them. As I have said, with respect to the goal of meaningful emissions reduction this cannot work in the current global economy. By the time we get around to efficiency with sacrifice, if indeed we ever do, the game will surely be up. We will not solve the climate crisis through efficiency measures alone because they leave the desire for more than we need untouched. Now, it is safe to say that most advocates of efficiency are more aware of the problem of climate change than some other members of the global prosperous (to say nothing of those outside this group). And yet the key similarity between the latter and the efficiency boosters is that neither appears willing to impose significant constraints on desire. To consume *less* rather than merely differently. David Pears has written that in cases of weakness 'desire quietly removes the intellectual obstacle to its own fulfillment' (Pears 1984, 19; my emphasis). This echoes Sparshott's way of parsing Aristotle's conception of weakness, according to which untrained desire diverts our attention from the reasons that would thwart it. Absent desire's psychic mischief, efficiency would surely be seen by its current advocates as the non-solution (or at best the half measure) it really is.[16]

Fortunately, our smaller part—the seat of our moral convictions—has potential psychic allies (beyond the imagination). As Woodruff puts the point, 'justice has to take the non-rational into account in order to be credible' (2011, 145). The forces he has in mind, and that are naturally allied to reason, are shame and honour as well as their subsidiary emotions, courage and anger. If these forces are properly cultivated, appropriate desires will follow. In the next section I will talk about all of them, focusing on shame and honour.

4.5 Modernizing Shame and Honour

Bruce Robbins argues that cosmopolitanism is an inherently paradoxical set of ideas because it asks us to strike a 'balance between attachment and detachment' (2012, 416). That is, cosmopolitanism is an impersonalist stance *on* the local and particular. One does not necessarily cease to belong to some particularity just because one is concerned to critique it in this way. The current trend among some political philosophers to defend a 'rooted cosmopolitanism' is an expression of this psychic and theoretical duality, but it also finds powerful expression in the phenomenon of shame.[17] In this section, I want to describe the general moral psychology of shame and honour, and see how these emotions might be modified or extended to apply to the climate crisis.[18] I begin with shame.

In determining what experiencing shame should involve, the best place to start is with an analysis of what it means for an agent to take responsibility for her wrongdoing. A key feature of taking responsibility is that an agent sees herself as an apt target of some moral community's reactive attitudes. This is where the comparison between shame and guilt becomes instructive. As has often been noted, guilt is an indispensable emotion for a genuinely repentant wrongdoer to experience because it directs her attention to what she has done (or failed to do) and, by extension, to her victim. By contrast, shame directs the wrongdoer's attention to herself.[19] A shamed agent sees herself as morally compromised by her exposed wrongdoing. Velleman is, moreover, surely correct to connect shame's unwanted exposure to a certain kind of anxiety: 'threats to your standing as a self-presenting creature are thus a source of deep anxiety, and anxiety about the threatened loss of standing is, in my view, what constitutes the emotion of shame' (2006, 55). Either emotion can be present without the other. Shame in this case is not tied to the intentional or to agential moral responsibility.[20] For example, one may be ashamed of one's family or physical appearance. Guilt, on the other hand, can be appropriate when an agent has done something morally wrong but legitimately judges that he need not make any deep negative judgements about himself as a result. Moral dilemmas are an interesting example of this.[21]

Although there are cases in which the two emotions operate independently of one another, there is a large domain of cases where they are found together. The relation between the two emotions in this domain is asymmetrical, with shame providing the normative lead. As Bernard Williams has argued, shame reveals an agent's links with an entire moral world, one of which he is a lapsed member and out of which his very identity as a moral agent is formed:

> [Guilt] can direct one towards those who have been wronged or damaged, and demand reparation in the name, simply, of what has happened to them. But it cannot by itself help one to understand one's relations to those happenings, or to rebuild the self that has done these things and the world in which that self has to live. Only shame can do that, because it embodies conceptions of what one is and how one relates to others. (1993, 94)

What is under siege in the shamed agent, insofar as she is undergoing a crisis of identity, is her agency. The shamed agent is not just noticing a fact about herself—as is perhaps suggested by the recurrent metaphor of an unclean or otherwise tainted soul—but instead finds that she is radically unsure about what to do in the moral sphere. She is exiled from her moral community without being excommunicated from it, a way of putting things that brings us back to my earlier point about shame's psychic duality. Although the experience of shame is often thought of as unambiguously negative, it is important to highlight this constructive aspect of it. To go through the process of shame is to rebuild one's agency in accordance with the better part of oneself. This is what makes shame interesting for the current discussion.

Further, as Appiah has shown, moral shame can manifest in collectives and can thus be a powerful force in moral revolutions: 'One day, people will find themselves thinking not just that an old practice was wrong and a new one right but that there was something shameful in the old ways. In the course of the transition, many will change what they do because they are shamed out of an old way of doing things' (2010, xvii). The four examples Appiah discusses are the duel, footbinding in China, the Atlantic slave trade, and the (ongoing) war against women in some Islamic societies. Those opposed to the practice in question made explicit appeal to how it looked to certain outsiders. This appeal was, however, complex because the 'progressives' did not thereby place themselves entirely outside their own culture. Rather, they would point both to a culturally external gaze and to a partially dormant intra-cultural gaze. Here, for instance, is what Appiah says about the Chinese practice of footbinding:

> In dealing with the problems facing their society at the end of the nineteenth century, China's modernizing intellectuals, like those who resisted them, were guided by a profound loyalty to their nation and to its deepest intellectual traditions. Many of the modernizers insisted on the distinction between *ti* (substance) and *yong* (application); they believed, as they said, in 'Chinese learning for fundamental principles, Western learning for the practical applications'. (2010, 87)

The external (in this case Western) ideals are used to leverage a certain self-conception of the collective which has become buried or distorted through the operation of a particular (shameful) practice. The appeal would likely be ineffective were either element missing. Had the 'modernizing intellectuals' simply pointed to putatively superior Western practices, they would have been attacked as decadents or even traitors; had they appealed simply to the old ways—footbinding was unknown in the time of Confucius, for example (Appiah 2010, 87)—they would have been derided as nostalgic cranks.

How might this apply to the case of climate change? In October, 2013, Typhoon Haiyan crashed into the Philippines, causing immense destruction. The next month, at the 19th UN Conference of Parties (COP) meetings in

Warsaw, Yeb Sano, the chief climate negotiator for the Philippines, made this emotional speech:

> In solidarity with my countrymen who are struggling to find food back home, I will now commence a voluntary fasting for the climate. This means I will voluntarily refrain from eating food during this COP, until a meaningful outcome is in sight ... What my country is going through as a result of this extreme climate event is madness, the climate crisis is madness. We can stop this madness right here in Warsaw. (Quoted in McGrath 2013)

Sano's presentation cuts through a lot of noise about the way climate change impacts real people. Assuming you are not a dyed-in-the-wool climate cynic, it is the sort of appeal that is capable of making you reassess where you stand and what you can do to change things. On hearing the speech you might be inclined to at least go out and look at the facts more carefully than perhaps you had hitherto. You might ask what could have prompted this respected figure to say and do these things. In other words, you might at least begin the process of internalizing this person's gaze. Yeb Sano was speaking explicitly on behalf of both his country and that of all the poor nations of the world who will have to adapt to climate disasters but who are (mostly) not themselves responsible for climate change.

But just as importantly he is implicitly giving voice to the concerns of people of the future. The same can be said of many others: James Hansen, Vandana Shiva, David Suzuki, Naomi Klein, Bill McKibben, Al Gore, Elizabeth Kolbert, George Monbiot, Mark Lynas, and more. I view such people as both moral exemplars and proxies for future generations. To take them seriously is effectively to internalize the gaze of future people. This can begin to make us feel ashamed by the conduct of the group—our economic class, nation, generation—to which we belong. But, again, the external gaze alone is not sufficient. Internal leverage is required too. Appiah notes that in all his historical cases the progressives or revolutionaries did not need to invent new moral principles in order to move the society forward. Here is the point applied to the slave trade: 'By the mid-eighteenth century ... both among the religious and the anti-religious, slavery was widely understood to be wrong. And so, as with the duel and with footbinding, what galvanizes the movement against slavery is not moral argument: the arguments are in place well before the movement begins' (Appiah 2010, 110). Similarly, throughout this book I have been arguing that in order to face the climate crisis squarely we do not require radically new moral principles. We *know* it is wrong to wreck the climate. Just as the opponents of footbinding could appeal to intra-cultural, though dormant, principles or traditions, so can we. The call for moral revolution implores us to live up to our best ideals, not to transcend them. For example, in 1948, the U.N. passed a resolution calling on member states to intervene anywhere a genocide was taking place or *likely to take place*. That document is part of our moral self-definition. I have already argued (in Chapter 3) that if those working on security and climate change are correct we should be very worried about the possibility of crimes of atrocity in the near future. That we are allowing this to happen should be a source of shame given our explicit collective decision to work at preventing such things from happening. As Michael Morgan puts it, there 'might be a shame at being human at all in a world in which there was an Auschwitz and in which there seems to be no reliable obstacles to its repetition' (2008, 26).

To advance the analysis we need to examine the connection between shame on the one hand and honour, anger, and courage on the other, and show that these emotions can be brought to the defence of specifically Enlightenment values. Like shame, honour has its indisputably ugly manifestations, but it has of late nevertheless enjoyed something of a renaissance among moral philosophers.[22] Part of the reason for this is its connection to justice. Other things equal, we want the just resolutions of our conflicts to last. But the only way this is going to happen is if the resolutions allow all participants to believe that the outcome is honourable, that they did not have to sacrifice anything of deep moral importance to them in the process of negotiating a deal. Otherwise, some participants will seek to undermine the settlement at the first opportunity, possibly damaging the interests of the collective in the process.

What makes an honourable settlement particularly difficult to achieve is that in all such situations there are going to be redistributions of benefits and burdens. The trick is to see the latter as something more than zero-sum reallocations of resources. As Woodruff puts it, this is the only way to ensure that the participants can continue to work together as a team after the settlement (2011, 157). As I will suggest at the end of this section, we should not overestimate the extent to which such cohesion is possible in the case of large collectives, but the general aim is laudable. Anthony Cunningham has recently provided an account of 'modern honour' that ties it to specifically Enlightenment values of liberty, fraternity, and equality. His key claim is that what many have taken to be an essential characteristic of honour is in fact an historically contingent accretion, though a culturally pervasive one. Honour, on this standard conception, is intrinsically about exclusion through competition and invidious comparison. And because they are matters of such deep personal and communal concern in many honour societies—having to do, for instance, with the protection of a female family member's sexual purity—the comparisons and negative evaluations that flow from them can lead to profound harms, like the 'honour killing' of a raped sister or daughter.

An honour system has three features. First, the specification of certain excellences. Second, the comparison among a select group of agents in terms of their success at achieving the excellences. Third, the distribution of rewards and punishments on the basis of these outcomes. The benefits of honour systems have to do with the emotion's psychological depth, the way it can motivate us to act because of its connection to our sense of who we most fundamentally are. This is why attacks on our honour are so important to avoid, and why we can get so angry in the face of them. The intersection of these emotions also defines the field of courage. We will generally be more inclined to face down fear if the threats we perceive touch our sense of honour. Because of all of this, honour is a superb social glue. Now, if the chief desideratum in addressing climate change is moving from Cripps' should-be collective to an actual collective, and if in order to accomplish this we must do battle against powerful contrary-to-virtue desires (those involved in excessive consumption), then it looks like we need the assistance of just this sort of non-rational psychic force.

The question is whether or not we can have an honour system without the kinds of invidious comparisons that result in zero-sum allocations of important social resources. Cunningham thinks we can: 'The gravest danger here comes in the form of systematically highlighting excellences that seem designed to uphold the privileges of some at the expense of others. Nothing about the concept of honor beckons us to do so' (2013, 58). Actually there are two questions here. The first has to do with how honour is compatible with rejecting the social privileges of the few. The second has to do with how we can develop a sense of honour for the climate crisis. I'll begin with the second question and come back to the first one at the end of this section. At the level of individual agency, how do we begin to conceive of our luxury emissions, for example, as dishonourable, rather than simply wrong? Unfortunately, I see no magic bullet here, though in some ways this is the most important question of all. But we can gain some traction on the issue by coming back to the notion of internalizing an outsider's gaze in the process of moral repair. In elaborating this process, we will see just how close shame and honour are as they function in the moral-psychological economy.

Imagine Marc, a Canadian citizen, watching his country make its usual mischief at COP meetings in an effort to undermine any meaningful global deal. Whereas he had always looked on these antics with a sort of resigned or rueful detachment, suppose Marc has his compassion stirred by Yeb Sano. As a result he now feels a creeping sense of anger at the Harper government's strategies. He believes the government is acting contrary to Canadian values of internationalism, openness, toleration, and equality. These values, he believes, describe certain Enlightenment excellences which his country had always tried to instantiate in its foreign policy. He might be inclined to act by voting for the Green Party in the next federal election, writing letters to the editor, mocking the Minister of the Environment to his friends at work, and so on. However, more disturbingly, Marc might also realize that his family's very high standard of living is tied to the perpetuation of the global fossil-fuel regime which his government is aggressively seeking to expand. It may be that he works in an industry with indirect ties to the Alberta tar sands or his pension fund might be heavily invested in fossil fuels. The upshot is that Marc now realizes that his *desires* are implicated in his government's wrongdoing. In attacking his government he is also exposing the minutiae of his own life to moral view. But for a long time he cannot bring himself to change the way he lives, so entrenched are his habits.

Marc will likely not acquire the fortitude he now requires all by himself. This is why exemplars are so important in times of moral crisis. As Aristotle saw, the point of an exemplar is to provide a model whose actions and thought patterns can be more or less *copied* by moral apprentices until the latter develop the ability for fully autonomous decision-making. Marc is an agent dishonoured by the actions of a wrongdoing group to which he belongs. Insofar as he judges himself to have been caught up in that wrongdoing, however indirectly or passively, he now questions his capacity for practical deliberation aimed at the good. The critique was relatively easy until he became aware of his own complicity, at which point he needed the external source, the spokespeople for the future, to help guide his thoughts, feelings, and actions. What exactly Marc should do next is an open question, but if his sense of honour is truly tested by what his government is doing we can assume that his actions—perhaps running for the Green Party rather than just voting for it, total personal divestment from fossil-fuel companies, civil disobedience, etc.—will be significantly riskier than what he had done before. The key point is that his desires are being reshaped in this process.

I'll come back to Marc's transformation in a moment, but at this point we might wonder about the difference between shame and honour. Since both emotions involve internalizing an external gaze, are we not dealing with a single moral-psychological phenomenon? Here, we need to make a distinction between the sense of honour and the fear of shame on the one hand and the experience of dishonour or shame on the other. The first pair describes features of the successful moral agent; the second what is involved in moral failure. To begin, we should note the secondary emotions or values that tend to accompany the sense of honour and the fear of shame, respectively. Both of these intentional states involve the disposition to live up to a communally specified ideal. However, the sense of honour does so with pride and admiration, stressing the nobility or beauty of the ideal, while fear of shame is an anxiety about the consequences of the failure to uphold the ideal. Every honourable person also fears shame's fall. The main difference has to do with the level of moral experience an agent has attained.

Since she can still be surprised by the world and herself, an agent in the earlier stages of the process of moral education will be dominated by the fear of shame, while to the extent that she has gained significant experience of the world, as well as of herself, the sense of honour will tend to dominate. Needless to say, there are many points along this continuum. While we cannot understand the nature of failure in any domain without knowing what success there looks like, climate change inaction is a clear moral failure, so we should focus on the experiences of shame and dishonour, which have to do with recognizing one's failure to live up to the extended cosmopolitan ideal. Here we are probably not dealing with as sharp a distinction as we were with the sense of honour and the fear of shame, but there is one promising basis for distinguishing them. We might say that the experience of shame looks mainly to the social exclusion consequent on the failure whereas the experience of dishonour looks mainly to the lost ideal or excellence. Again, both will be operative in the moral psychology of real agents, but we can distinguish them analytically.

With these distinctions in hand, let's come back to Marc. Since he has internalized broadly Enlightenment values, Marc is certainly no moral neophyte. At the beginning of his story his commitment to these values is, however, merely mind-deep. His most important insight is that many of his desires are incompatible with his ideals, i.e., that he is a quintessentially morally weak agent. Spurred by the experience of shame/dishonour and aided by courage and anger he transforms his desires and the choices and actions consequent on them. He moves from weakness to strength. In the process, he sometimes trains his moral gaze on the human community from which he now feels exiled—the one encompassing people of the future—while at other times he focuses on the ideal of radical equality which his group has dishonoured. In this way he shuttles between the experiences of shame and dishonour. So Cunningham is right to insist that there is a distinctively modern way to think about honour: we can look to the defence of broadly Enlightenment ideals as a matter of honour and to our failure to uphold these ideals as a matter of dishonour. Further, the approach to moral failure examined here vindicates the focus on moral weakness rather than wickedness, at least with respect to the climate crisis. Many more of us are like Marc than David Koch.

Finally, let's return to the first question I posed about Cunningham's analysis. Can there be a universally inclusive honour system or is that notion actually oxymoronic? In other words, won't a commitment to honouring Enlightenment values result in the social 'exclusion' of those who reject this moral focus? Cunningham is probably too sanguine about this. Much hangs on how we understand excellences and the social privileges that flow from achieving them. I have been stressing in this chapter and the previous one that we cannot meet the demands of intergenerational justice without significant scaling back, and that this will necessarily involve some pain on our part. The wholesale effort to decarbonize the global economy is undoubtedly going to produce winners and losers. The excellence of the relevant subset of our institutions and individual actions will be measured by the degree to which they genuinely strive toward protecting the vulnerable. It is difficult to see how this ideal can be achieved or effectively maintained unless our social system also honours those who succeed, or strive sincerely to do so. They are the ones who not only will but also should be put in positions of power and prestige, and this is bound to breed some social friction.

Fossil-fuel-company executives whose firms continue to dig for coal, shale oil, and tar sand oil and who both lobby national governments for subsidies and aggressively oppose large-scale development of green energy technology are *enemies* of the future. It is probably a mistake to think of all such agents as only contingently or shallowly committed to their aggressively extractive stance towards nature. Like the Koch brothers, many of them think it *morally important* that the age of fossil fuels be perpetuated for as long as possible.[23] These people are the latest manifestation of a very old attitude according to which unfettered exploitation of the earth's resources is an honourable, even divinely sanctioned enterprise. It is our *job* as humans, and a vital expression of our freedom, to transform the natural world technologically while taking little heed of principles of preservation or environmental justice.[24] The principle of substitution endorsed by so many of our economists—the claim that no resource is really finite since its place in our 'consumption bundles' can be filled by something equivalent if we run out of it—tells us that nature is literally inexhaustible. We should take what we can get without worrying that our actions might undermine the welfare and freedom of future people.[25]

Ideally, such people—as well as the politicians, media pundits, and citizens who support them—can find a way of reconfiguring their sense of who they are such that the principle of intergenerational justice can take hold in their souls. This brings us back to Woodruff's ideal of honourable settlement which as such allows collectives to move forward. The upshot of the present discussion is that we should not be naïve about this possibility. Some vested interests appear to be too firmly entrenched for it to happen. At least with respect to the climate crisis, I conclude that Cunningham is probably incorrect to claim that we can appeal to modern honour in a way that is not to some extent dependent on a system of privileges that involves exclusion. But this should not trouble us overmuch since the proposed system is morally justified in virtue of explicitly incorporating the vital interests of future people, and because if we are to move humanity in this direction we require honour's (and shame's) power to unify us both as individuals and as collectives.

4.6 Conclusion

Psychic integration is always going to involve something of a battle between desire and reason for the middle part of the soul (the house of shame and honour). Some degree of inner moral strife is inevitable for beings condemned to a choice between weakness and (mere) strength. Wholeheartedness does not imply absolute harmony—only full virtue can achieve that—for there will likely always be rebellious desires in us. As I have argued, weakness works in us by partially blocking our access to our ideals, or their inferential connections. As a rule weak agents turn readily to the evasions of rationalization and self-deception. This implies that the problems we face are epistemic as well as moral. Moral weakness is often enabled by self-deception. This is why I quoted Pears, in Section 4.3 above, to the effect that in the weak, desire removes the *intellectual* obstacle to its own fulfilment. Aided by the wayward desires I have already mentioned, our smoker gets herself to believe that she's got the habit under control (though she's smoking more lately); our couch potato that he would probably have injured himself at the gym (though this has never happened before); and our activist that her candidate doesn't have a chance anyway (though the latest polls suggest otherwise). Facing the climate crisis squarely is indeed as much about how we ought to organize our beliefs as it is about how we ought to act, and it is therefore to an analysis of these issues that we now turn.

5

Truthfulness

5.1 Introduction

What do a bureaucrat who edits official government reports about climate change so as to emphasize our 'uncertainty' about it to the public (in the teeth of a growing convergence among scientists about its reality and urgency), a U.S. Senator who declares that scientists have conspired to fool the public about the scope of climate change, a statistician who claims that we cannot afford to reduce our carbon emissions and should adopt a wait-and-see approach to the problem, and a resident of a tiny village who counts herself an environmentalist but whose prosperity comes from her country's exploitation of fossil-fuel resources, have in common? The answer is that they are all in denial about climate change. But as this diversity of examples illustrates, climate change denial has many faces and it is crucial to explore the differences among them. Climate change denial is a highly corrosive epistemic habit in our culture. Titania, from *A Midsummer Night's Dream*, captures the problem nicely:

> The human mortals want their winter here;
> No night is now with hymn or carol blest...
> Diseases do abound.
> And through this distemperature we see
> The seasons alter...change
> Their wonted liveries; and the mazèd world,
> By their increase, now knows not which is which.
> And this same progeny of evil comes
> From our debate, from our dissension.
> We are their parents and original (Shakespeare 1995, 2.1, 105–15)

Climate change denial is a degraded form of 'debate' or 'dissension' that has brought us a 'progeny of evil' and will continue to do so until we grasp its complex meaning more fully. This chapter seeks to advance our understanding in this area by examining the phenomenon of denial in the context of the epistemic virtues and vices.

I begin by arguing that we need the explanatory tool of the intellectual virtues and vices to make full sense of what has gone wrong with our beliefs about climate change. To this end, I offer a picture of Aldo Leopold and James Lovelock as ideal epistemic experts. But this is mainly to get a general picture of how the epistemic virtues might be applied and to provide a contrast with the first three (out of four) faces of climate change denial. Since most of us are non-experts about climate science, but still need to structure our beliefs about the latter in a principled way, I move next to an analysis of the epistemic status of the IPCC, our chief source of information on the physical science of climate change. Employing the terminology of Roberts and Wood, I characterize this body as a 'hetero-regulator' of our beliefs and show that adopting the beliefs it prescribes is (qualifiedly) justified. Next, I examine the fourth face of climate denial, self-deceptive denial. I argue that our self-deception is the product of what Barnes calls an anxiety-reducing bias. Finally, I lean on the psychoanalytic literature on climate change to show that while our anxiety-induced self-deception is surely a form of 'negation', the real danger is that it might develop into a more pathological 'disavowal' of the problem.

5.2 Three Faces of Climate Change Denial

Let's begin by looking at three distinct kinds of climate change denial. We tend to overlook the differences among these three types but, even though I will ultimately be emphasizing what they have in common, it is important to note the dissimilarities. The first is *temporizing denial*, which may not look like denial at all. For example, although he claims to believe in the science of climate change (and there is no reason to doubt him on this score), Bjørn Lomborg's views have become important to those who want to perpetuate economic and political business-as-usual while paying lip service to the problem of climate change. Like many of these people—the accelerators— Lomborg (2001, 2010) thinks we should get as rich as we can now and use our wealth to combat the effects of climate change later.[1] Unlike the next two types of denier, this one denies the urgency of the crisis rather than the facts themselves, and so it can make its proponent look science-friendly and even progressive. Wilfrid Beckerman (1992), another advocate of this approach, implies clearly that mitigation is a waste of resources. Investment in growth industries would benefit people of the future far more than reducing global GDP by curtailing our use of fossil fuels.[2] But it is untenable to suppose that the urgency of the crisis we are in is not itself one of the facts of climate change. Belief in climate change, as we have seen throughout this study, involves, inseparably, both the affirmation of certain facts about the world and the recognition that those facts demand swift action from us, aimed in the first place at mitigating the future impacts of climate change (through rapid decarbonization).

The second form of denial is certainly the crudest since it involves the assertion of deeply implausible claims about climate change. Call this *fabulous denial* ('fabulous' in the sense of 'incredible' or 'exaggerated' rather than 'terrific'). The climate change fabulist comes in at least two varieties. First are those who claim that climate change is a theoretical 'hoax' cooked up by climate scientists conspiring to hoodwink the rest of us. There has been little attempt to explain what might be motivating scientists to do this, beyond all the grant money they are allegedly so hungry to secure, and not a shred of evidence has been presented to back up the claim. James Inhofe, the Republican Senator from Oklahoma, is the most notorious example of this sort of thinking. He continues to advance fully refuted 'scientific' claims in order to push the idea that climate science, as presented to us most prominently by the IPCC, is a politically motivated grand illusion. Of course he has been well funded by the oil and gas industry, having received campaign contributions totaling $1,414,996 over the years (Hoggan 2009, 96).

One need not deny anthropogenic climate change in this fashion or impugn the motives of scientists in order to qualify as a fabulist. Another way to earn the label is to insist that climate change is going to alter the planet for the good of all. For example, in 1992 the Information Council on the Environment (ICE)—which, just like Inhofe, is funded by the fossil-fuel industry, in this case the Western Fuels Association (Vanderheiden 2008, 31)—released a video called *The Greening of Planet Earth*, meant to celebrate the carbon-rich future of our planet. Gelbspan characterizes it thus:

> In near-evangelical tones, it promises that a new age of agricultural abundance will result from the doubling of the atmospheric concentration of carbon dioxide. It shows plant biologists predicting that yields of soybeans, cotton, wheat and other crops will increase by 30 to 60 percent—enough to feed and clothe the earth's expanding population. The video portrays a world where vast areas of deserts are replaced by grasslands, where today's grass- and scrublands are transformed by a new cover of bushes and trees, and where today's diminishing forests are replenished by new growth as a result of a nourishing atmosphere of enhanced carbon dioxide. (Quoted in Vanderheiden 2008, 31)

This verdant utopia offers tremendous benefits for humans but we can realize it only if we rigorously avoid regulating fossil fuels, especially coal. We can finally put to rest the twin worries of overpopulation and climate change, since the latter is precisely our solution to the former. Now, there's no need to deny that in the short-term we could see some benefits for humans from increasing temperatures—a reduction in cold-related mortalities, for example—but, as the IPCC makes clear, 'the balance of impacts' on humans from climate change 'will be overwhelmingly negative' (Quoted in Friel 2011 87).

Just because they are so difficult for any minimally informed person to believe, the two kinds of fabulous thinking just sketched do not represent a deep threat to our epistemic well-being (although *The Greening of Planet Earth* was evidently much admired by officials in the George Bush II administration). The same cannot be said about the third face of denial. Here I have in mind the sort of 'manufactured uncertainty' or 'contrived skepticism' (Vanderheiden 2008, 33) that has become so ubiquitous in our public discourse about climate change. The explicit goal of these deniers has been to sow confusion about the issue so as to perpetuate policy inertia. This is the same strategy employed by cigarette company executives, who proclaimed that 'doubt' about the links between second-hand smoke and lung cancer was their 'product' (Hoggan 2009, 64–7). To deny climate change in this way is to use one's privileged epistemic position to spread doubt, confusion, or outright falsity about an issue of deep public concern. This is *bullshitting denial*.

As Frankfurt has argued, the problem with the bullshitter is her indifference to the truth. The liar must track the truth in order to oppose it while the bullshitter need look only to what must be said in some context to gain strategic advantage there. But to accomplish this she must tap into our expectations regarding how truth presents itself. Since in our culture science has a certain epistemic prestige, the bullshitter must appear to be its friend. This is why Frankfurt says that the bullshitter attempts to deceive us not about 'the facts, nor even about what he takes the facts to be', but about 'his enterprise' (1988, 130). The latter has nothing to do with revealing or discovering the truth, and *this* is therefore what she must hide about it. By contrast the liar needn't pretend to care about the truth because, in a warped way, she does care about it.

To see how bullshit typically works think of the process of cultural colonization. At its most nefarious this takes place not by brute force alone (though this, or its threat, is often a handy supplement) but also through the colonizer's ability to manipulate the language and transform the rituals of the colonized. The new beliefs then spread from within the body of the culture itself, like a virus. In this sense the original Jesuit missionaries were consummate bullshitters. In *The Orenda* Canadian novelist Joseph Boyden shows us how this worked in the first contact between French Jesuits and the Huron in seventeenth-century New France. The Jesuits feigned respect for certain carefully targeted Huron beliefs, practices, and rituals, all the while seeking to infuse these with Christian content. As I see it, the process of adhering to the form of a certain practice or set of practices—including, crucially, their linguistically expressed repertoires of norms, values, and procedures—while hollowing out their contents is the essence of bullshit. With respect to our issue, the phenomenon shows up in the creation of sundry fossil-fuel industry supported think tanks, coalitions, and institutes, all devoted to churning out ideologically charged denunciations of climate science; the alteration by bureaucrats of political documents to obscure the truth about climate change; the intimidation of climate scientists by politicians;[3] the publication in seemingly legitimate journals—like *World Climate Review*—of mostly unsupported and industry-friendly 'scientific' claims, and more.[4]

More particular examples abound, but none may be more telling than the 'sound science crusade' of certain members of the U.S. Congress. In 1995, the House Science Subcommittee on Energy and the Environment held a number of hearings whose goal, in the name of promoting sound science, was to undermine progressive environmental legislation. The process was known as 'Scientific Integrity and the Public Trust.' Vanderheiden summarizes it thus:

> Describing climate change as based on 'junk science'—a favorite term of anti-environmental ideologues, used in reference to any credible scientific research that finds domestic industry or consumption patterns to cause adverse environmental consequences—the central strategy of these hearings was to call into question the integrity of mainstream science, and thereby to discredit the empirical bases of the nation's environmental policies. (2008, 37)

The idea is to use the prestige of scientific rationality—especially the scientific commitment to caution and scepticism—against science's content as expressed, in this case, in propositions widely accepted by climate scientists themselves. As this example illustrates, Frankfurt is surely correct to argue that bullshitting is worse than lying because, both as consumers and producers of it, it 'unfits' us for the truth (1988, 132). Under assault from bullshit truth loses its epistemic authority. In place of propositions backed by evidence we get propositions repeated ad nauseam, the very repetition serving to crowd out alternative propositions. This is why it helps the bullshitter immensely to control the field of communication as much as possible. In the face of this onslaught, social space for the truth shrinks, and would-be knowers have a predictably difficult time finding it. The analogy with the process of cultural colonization is thus apt: the bullshitter's ultimate aim is to colonize epistemic space, the better to control the development of public policy.

Why think about climate change denial—that of the temporizer, the fabulist, or the bullshitter—from the standpoint of the intellectual virtues rather than more traditional epistemological themes of truth, justification, and warrant? For example, is it not sufficient to criticize Inhofe on the grounds that his views are manifestly false? The matter seems relatively straightforward. Inhofe made his infamous remarks to Congress in 2003, claiming to be bringing to wide public attention scientific truths 'that refute the anthropogenic theory of catastrophic global warming' (2011, 169). He went on to discuss, at some length, the history of IPCC reports, the accuracy of climate modelling, the reliability of satellite and weather balloon data, information about the status of the world's glaciers, the health risks of climate change, and much more. All the claims he made had been shown to be groundless at least five years before, with the publication of Ross Gelbspan's *The Heat is On*.[5] So why not say that the views he espoused are unjustified and move on? The answer is that this assessment of his epistemic behaviour does not go far enough. To explain this reaction, let's begin by talking about the virtues of epistemic exemplars in this domain. Anthropogenic climate change is just one aspect of a broader phenomenon concerning the nature of planetary systems, or *the* planetary system, and how humans fit into them (or it). So in what follows I'm going to examine the central ideas—and also, briefly, the professional practices—of James Lovelock and Aldo Leopold, thinkers whose work is focused on this broader phenomenon. With a description of these agents in place, we can return at the end of this section to the three faces of denial.

As we seek out worthwhile objects of knowledge, two criteria stand out: the object's intrinsic importance and the contribution knowledge of it makes to flourishing.[6] For example, it is good to want to know about the way carbon atoms move over the millennia through planetary systems.[7] There's beauty and complexity in the carbon cycle, but also an underlying simplicity in its sheer repetition, expressible in our theories about how the whole thing works. These are the defining qualities of intrinsic importance. This can all be appreciated apart from any practical use we might make of the knowledge. However, knowing about many intrinsically important objects can also contribute to our flourishing.[8] This is certainly true of our knowledge of the carbon cycle. It is both beautifully complex and, since we are now interfering with it in a way that is bringing about a crisis, important for us to know about if we are to flourish as a species. When our concerns with both flourishing and intrinsic importance come together this way, the objects picked out are maximally worthwhile. On this understanding, whatever else it is, anthropogenic climate change is a truly awesome potential object of knowledge. We know much more about it than we did a hundred years ago, but there is much more to learn and we have barely begun the task of relating what we do know to our understanding of human flourishing.

Since the publication of *Limits to Growth* in 1972, we have learned an extraordinary amount about such systems. Lovelock and Leopold have both stressed the need to know more, and to think and feel differently, about the way we humans properly fit in to larger systems. Consider the *Amsterdam Declaration on Earth System Science* (IGBP 2001) put out jointly by four international climate change research programmes. It attempts to codify Gaia theory as legitimate science and comprises five essential claims. The first is that the 'Earth system behaves as a single, self-regulating system comprised of [*sic*] physical, chemical, biological, and human components'. Second, anthropogenic interference in the components of this system is now real and discernible. This shows up in the gaseous composition of the atmosphere, biological diversity, the water cycle, and biogeochemical cycles, among other things, all of which are undergoing changes beyond those attributable to natural variability. Third, global change cannot be understood in simple cause–effect terms but only though the concepts of feedbacks, cascading causes, 'multidimensional patterns', and so on. Fourth, the Earth system undergoes abrupt changes, often on a decadal scale. Finally, the Earth is 'now operating in a no-analogue state': with respect to key parameters the system is outside the natural variability operative for at least 500,000 years.[9]

What could it mean to take these insights seriously in the process of belief formation? Think in this context of the much-discussed epistemic transformation of Aldo Leopold from forest manager to ecologist. Initially, Leopold's task was to ensure that the forest remained viable for economic purposes by keeping the deer population robust for hunters. This involved shooting wolves on a large scale (among other things). But Leopold was surprised to discover that the decline in the wolf population was causing a boom in the deer population and that this was having negative knock-on effects on vegetation. Leopold's transformation is twofold. First, his call to 'think like a mountain' expresses the need to think in systems. Second, he reconceptualizes the role of our species in these systems, 'from conqueror of the land-community to plain members and citizen of it' (1986, 327–64).

Although they don't put the point this way, Hirsch and Norton effectively synthesize the insights of Lovelock and Leopold in claiming that we need to 'think like a planet'. More importantly, the transformation they envision is clearly an epistemic one: 'Leopold's account provides an important case in how a conscientious individual, armed with the latest science, can undergo a transformation in the mental models that shape his or her reasoning' (Hirsch and Norton 2012, 320). This strikes me as correct, but Hirsch and Norton go on to argue that Leopold's transformation should be viewed less as the product of a scientific discovery and more as a change in the key metaphors that shaped his view of the wild. For the climate crisis, we too need to move beyond the 'productivity metaphor' and towards more appropriate, systems-based metaphors. I see no need to deny the importance of metaphor in our thinking, but the interpretation of Leopold is forced. To be sure, Leopold *did* talk about the need to change our affections along with our beliefs and behaviours, but that, I suggest, is because his appreciation of natural systems also made him keenly aware of psychic systems: the way emotions, motivations, beliefs, and actions can work synergistically. But all of this is compatible with the idea that his original epiphany was in large part an epistemic one. He saw that the

Pinchott-esque, or conservationist, view of the mountain and its inhabitants was false to the reality he was encountering. While Gaia theory, as codified in the Amsterdam Declaration, provides propositions worthy of belief, Leopold gives us clues about how such propositions can be folded into one's larger epistemic and behavioural economy.

I see Leopold and Lovelock as important epistemic exemplars for our times, but in a fairly abstract sense. In their insistence on putting forward versions of natural systems analysis well before the current scientific vogue for doing so, they display immense intellectual courage and fortitude in the face of professional resistance. More importantly for my purposes, each thinker strikes an ideal balance between autonomy and heteronomy—that is, independence and dependence in the belief acquisition process—and this is the 'abstract' feature of their epistemic dispositions that I think is worthy of our emulation. The autonomy is obvious since each worked on the fringes of his intellectual community, but in taking seriously the insight that we ought to think like plain members and citizens of the biotic community rather than its lords and masters (this is Leopoldian language, of course, but the *Revenge of Gaia* contains the same kind of thinking), they display an intellectual humility that is still difficult for many of us to adopt. Humility is the recognition of dependency and to recognize or be dependent is to be heteronomous. This is as true in the domain of belief as of behaviour. The question is whether or not there is a virtuous form or degree of dependence in either domain. I will return to this question in Section 5.3 but want to close this section by looking at how the analysis just presented can enhance our understanding of climate change denial.

Climate deniers clearly lack truthfulness, but what exactly is truthfulness? Three elements stand out. First, and most importantly, it involves what Montmarquet (1987) calls 'epistemic conscientiousness'. This has to do with an agent's motivation: the epistemically conscientious agent has a motive or desire to arrive at the truth in her deliberations and to disseminate it as is fitting. Second, in their epistemic deliberations, truthful agents employ the best available domain-specific truth-tracking methods, taking into account their own abilities to make use of those methods and seeking out ways to cultivate these abilities appropriately. Third, they avoid spending an inordinate amount of time and energy on trivial and therefore unworthy objects of knowledge, especially where this focus interrupts the investigation of worthier objects. The complex disposition of truthfulness thus has to do with motive, skill, and focus. All of these can be developed by agents and so the virtue specifies an ideal that we can fall short of in a number of ways, thus rendering ourselves fit objects of moral condemnation to one degree or another. To generalize, we might say that it is not so bad to have the motive and the right focus while lacking the skill (unless one is culpable for having failed to develop the latter), worse to have the motive and the skill trained on trivial objects, and worse yet to lack the motive, whether one has the skill and proper focus or not.

Much climate change denial clearly involves the application of domain-specific skill—think again of Inhofe's painstaking elaboration of his claims or the trained bullshitter's often clever appropriation of the cultural tools of scientific communication—and we have just seen that the object of knowledge it picks out is eminently worthwhile, but the denier lacks the conscientious motive. This, I maintain, is the worst thing we can say about an epistemic agent. And, again, it comes out most clearly in climate change bullshit. Frankfurt's main point is that well beyond its original application, bullshit has the potential to undermine the whole enterprise of open and knowledge-based reason giving that is so central to a functioning democracy. These agents are therefore guilty of what Jason Baehr (2011) calls 'epistemic malevolence', a familiar example of which is O'Brien from Orwell's *1984*. The epistemically malevolent set themselves against the 'epistemic well-being' of other agents and this is, at bottom, what we need to say about all climate deniers examined so far, but above all the bullshitters. They seek to corrupt the minds of others and this is what makes them both especially reprehensible and qualitatively distinct from self-deceived deniers (examined in Section 5.4, below). If I'm right about this we have answered the question posed earlier about why we need the language of virtue and vice to describe climate change denial adequately.[10] It is an egregious epistemic failing which cannot be fully illuminated via the simple descriptions 'unwarranted' and 'unjustified' as applied to the cluster of beliefs expressing it. All three types of denial examined so far represent a failure to grasp how we fit into Earth's systems and the extent to which our discernable perturbation of those systems threatens to ruin our common future.

It is, moreover, salient that the examples of denial just canvassed involve those occupying positions of power and prestige in society. Though few deniers are themselves scientific experts, as heads of corporations, public intellectuals, or politicians they have a unique duty to come to grips with the insights of experts like Leopold and Lovelock at this seminal point in the history of our civilization. But to claim, as they all do one way or another, that we can continue with economic business-as-usual is to fail in this task. It is to assert instead an ignorant and arrogant superiority over the biogeophysical realities that ought to define the parameters of human economies. One does not need to occupy the sorts of positions just enumerated in order to qualify as a denier, of course, but I think that a significant part of the cultural phenomenon that is climate denial starts with these agents and those like them, from there trickling down to an army of unskilled or indifferent epistemic agents.[11] These forms of climate change denial are thus above all a failure of moral and epistemic leadership.

The main burden of this section has been to suggest that no explanation of this failure will be complete without invoking the role played in belief formation by bad epistemic character, not just arrogance but also the greedy desire for money and power. Some have suggested that the Enlightenment ideal of knowledge is ultimately to blame for all of this. According to Clive Hamilton, for example, the Enlightenment unleashes the 'autocratic subject', an agent capable of extracting truths from nature through the application of method but also free to ignore those truths when it suits her interests (2013, 29). Hamilton is wrong to claim that epistemic autocracy represents the Enlightenment ideal, rather than its perversion. I'll come back to this point in Section 5.3, below. But his critique of the autocratic subject certainly fits our analysis so far: climate change denial has shown us the ways in which greed and arrogance produce a dangerous illusion of autocracy and independence. Insofar as their way of seeing the world and our place in it is fuelled by humility and recognition of the need for constraint Leopold and Lovelock provide the salient contrast.

5.3 Hetero-Regulation: The IPCC

How can two scientific mavericks be genuine exemplars for the rest of us (or even for non-maverick scientists)? To begin, let's stipulate that a necessary condition of our beliefs being justified is that they are produced by sound reasons rather than non-cognitive forces.[12] The most obvious sources of information about climate change are the reports of the IPCC, but what does it mean to say that we have sound reasons to believe what these reports are telling us? The question is difficult because the complexity of these reports qua knowledge sources—that is, quite apart from their actual contents—is not sufficiently appreciated. They are (a) descriptions of the physical world produced by scientists (b) working in groups, (c) across cultures, and (d) in conjunction with political leaders. In turn, the reports are meant for (e) the general public as well as (f) policymakers, and are (g) usually filtered through the media. The IPCC reports are thus part of a nexus of social information that is bewilderingly complex as an object of study for the epistemologist.[13] How should we respond to this complexity? In this section I will look at three ways of doing so.

The first is relativist retreat. One might argue that because the work of the IPCC is refracted through so many competing social lenses, none of which has any special authority, we cannot make our epistemic way back to the source itself, the physical climate. Therefore, we should embrace epistemological relativism about what climate change ultimately 'means'. In the grip of this thought, we might come to believe that climate change is not primarily about the physical climate at all. It is instructive in this regard to look at Mike Hulme's *Why We Disagree about Climate Change*, both because the book has been enormously popular and because Hulme ultimately advocates this kind of relativism. Hulme insists that we should refrain from 'framing climate as overtly physical and global' (2009, 28). This is because our having come to think of climate in largely physical terms has allowed the idea to attain a 'near infinite plasticity'. Instead, we should, says Hulme, accept that 'the ultimate significance of climate change is ideological and symbolic rather than physical and substantive' (2009, 329).

For example, Hulme lists the views of deniers, catastrophists, gung-ho engineers and cynics as four 'myths' (in the non-pejorative sense) about climate change, all of which seem to be on an epistemic par (2009, 188–90). But I have argued in Chapter 3 that, suitably modified, the catastrophist position is the *right one to take*, and confidence about this can be rooted in the physical science as well as our best estimates about near-term future energy use as laid out by the IEA. So the claim of epistemic parity among these forms is unconvincing on its face. But the problems with Hulme's analysis go deeper. One of Hulme's key objects of focus is the problem of risk, which he believes is an intrinsically subjective concept. Because we all have different subjective assessments of what risks climate change poses, there is no Truth of the matter. But where significant harms are involved this is not persuasive. If because of severe drought I am forced to move my family to a place whose inhabitants are known to be hostile to members of my group, and that there is a non-negligible chance that we will be attacked and killed by those people, someone else would be either weirdly insensible or self-deceived to fail to appreciate the objective riskiness of my situation.[14]

If we deny that there is anything objectively risky about climate change, we might be led to the conclusion that it makes no sense to think about the phenomenon as a 'problem' at all. Hulme thinks that problematizing climate change has 'created a political logjam of gigantic proportions, one that is not only insoluble, but one that is beyond our comprehension' (2009, 333). I think it is correct to suggest that we have created a political behemoth here, but how is refusal to talk about a problem supposed to help us? The worry with Hulme's analysis is that, having rejected all examples of what it might mean to 'solve the problem' of climate change, we are left with vaguely spiritualized prescriptions (for example, we should adopt a religious perspective because it allows us to 'confront' issues rather than 'solve problems' (2009, 359–60)).

Hulme argues that we need to 'rethink how we take forward our political, social, economic and personal projects over the decades to come' (2009, 362). But what could define this forward trajectory if there is no problem to solve, if everyone's definition of danger, risk, and even 'the climate' itself is on an epistemic par? The way to resolve wicked problems like climate change is surely not to retreat to an anything-goes epistemic relativism, buttressed by a flight to the transcendent, but to take the science of climate change as authoritative and begin the hard work of prioritizing the many problems with which it confronts us. This is why I have been insisting throughout this book on our prioritizing the rapid decarbonization of the global economy. The huge success of Hulme's book is telling: for a culture that appears determined to do nothing substantial about climate change while pretending to take it seriously he provides the perfect cover.

The second way of responding to the complexity of the issue is to jump to the other extreme and insist that the IPCC ought to enjoy unquestioned authority, to take IPCC projections 'almost as if written in stone, like the message Moses brought down from the mountain', as Lovelock puts it (2014, 93). But this response, though better than the previous one, is also inadequate. The main reason for this is that the IPCC clearly makes mistakes, often significant ones. As Joe Romm remarks about AR5, 'like every IPCC report, it is an instantly out-of-date snapshot that lowballs future warming because it continues to ignore large parts of the recent literature and omit what it can't model' (18 August 2013). These are two rather major flaws. About the first one—instant obsolescence—consider these dates about the 2007 report. To be included in that report papers had to be written by the end of 2005. But given the lengthy process of peer review that any paper has to go through, even a paper submitted that late is probably working with data from 2003, at the latest (Dyer 2008, 93). Since climate science is advancing so rapidly, four years can make a very big difference in the analysis of a particular phenomenon. As for the second problem—'lowballing future warming'—we have already seen in Chapter 3 that IPCC reports have consistently underestimated key developments in natural systems, such as the rate of arctic sea ice decline, extent of sea level rise, rate of CO_2 emissions (especially in China), loss of Northern Hemisphere snow cover, and rate of permafrost melt.

Clearly what is needed is a middle position, a way of responding to the epistemic complexity of climate change, and the role of the IPCC in disseminating knowledge about it, that neither places all interpretations of the phenomenon on a par nor elevates the IPCC to the status of an unquestioned authority. Think of all of this from the standpoint of individual epistemic agents. What, or who, should we non-experts believe about climate change? Almost all of us face irreducible epistemic dependence here. To reprise a theme from Section 5.2, one way to approach the question has to do with striking a balance between the demands of epistemic autonomy on the one hand and the legitimate claims of external or expert epistemic sources on the other. The problem with criticisms of the Enlightenment like Hamilton's is that they throw the baby out with the bathwater. Surely at this point in our cultural history we are simply not in a position to reject wholesale the ideal of epistemic autonomy, the claim that individuals should to a significant extent be left to themselves to judge matters of truth and falsity. Do critics of the Enlightenment really expect us to defer meekly to the various experts, the same way our forebears were expected to submit their conscience to ecclesiastical authorities?

Though they do not talk about the IPCC, Roberts and Wood introduce the notion of a 'hetero-regulator', which can help us work through the complexities here. Hetero-regulation is meant to be a mean between extreme degrees of dependence and independence. Guided by the ideal of Leopold and Lovelock, this looks to be exactly what we are looking for:

> Clearly, an intellectual agent is wanted who has been, and continues to be, *properly* regulated by others, but has at the same time a mind of his own, being an independent and creative thinker and inquirer. If autonomy is to be a virtue, it must incorporate elements of intellectual heteronomy—not the vice, of course, but the phenomenon of being regulated by others. (2007, 260)[15]

This raises the question of the nature and justifiability of testimony, specifically of expert to non-expert testimony.[16] Although Hamilton implies that the Enlightenment's 'autocratic subject' will have no truck with testimonial beliefs, one of the key figures of the Scottish Enlightenment, Thomas Reid, places testimony at the heart of his epistemology. More importantly Reid thinks we require a balance between independence (the voice of 'Reason') and testimony:

> And, as in many instances, Reason even in her maturity, borrows aid from testimony, so in others she mutually gives aid to it, and strengthens its authority. For as we find good reasons to reject testimony in some cases, so in others we find good reason to rely upon it with perfect security, in our most important concerns. The character, the number, and the disinterestedness of witnesses, the impossibility of collusion, and the incredibility of their concurring in their testimony without collusion, may give an irresistible strength to testimony, compared to which its native and intrinsic authority is very inconsiderable. (Quoted in Coady 1992, 124)

We cannot help but base a significant subset of our beliefs on the testimony of others: when we were born, the elevation of a mountain we have never climbed, the safety of the food we eat, the results of last night's hockey games, the soundness of the doctor's advice, the report of the weather channel about today's likely rainfall, the belief that influenza is a virus, that grizzly bears hibernate for the winter, William is the Duke of Cambridge, Fellini directed *8 ½*, and so on. It would be difficult to go about our daily business at all in the absence of confidence in the truth of many such beliefs. Reid is insightful because he does not leave the question of belief justification up to Reason or testimony alone. It is well beyond the scope of the present inquiry to explore the general epistemological implications of this claim, but I do want to note the special way in which Reid has Reason strengthening testimony because it is relevant to the proper lay reception of IPCC reports. Let's focus on two ways reason does this.

The first has to do with scientific 'concurrence' and the related topic of possible 'collusion' among scientists. One of the reasons it is so implausible to claim that climate scientists are wrong, or just fabricating results, is the amazing agreement among various independent data sources with regard to the fundamental facts about a warming planet: tree-rings, sediment cores, ice-cores, the instrumental temperature record, corals, paleobotanical findings, computer models, anecdotal evidence, and more. For example, in a recent study scientists compared the instrumental temperature record to data from over 170 temperature-sensitive paleo proxies. The results showed remarkable concurrence among the sources, all of them indicating 'a significant warming trend from 1880–1995' (Armstrong, Mauk, et al. 2013).[17] This vast catalogue of independent cross-corroboration of basic data renders incredible the notion that scientists have colluded to falsify the facts. Let's call the scientists who put together the Assessment Reports 'witnesses' to the truth about the physical climate, or—since the IPCC synthesizes research without conducting it—as witnesses to the best research about the climate. Since 97 per cent of these witnesses affirm the fundamental facts about anthropogenic climate change, and collusion among them is ruled out (as we have seen), then, according to Reid, the testimony is likely reliable.

Second, consider the 'number and disinterestedness of witnesses'. One of the fundamental expressions of the virtue of disinterestedness displayed in ideal practices of science is that of peer review. The method of peer review expresses the virtue of openness to public reasons. This process is ramped up significantly in the IPCC's reports. For example, the 2007 Fourth Assessment Report (AR4) was over 3000 pages long. It incorporated the data and analyses of some 10,000 already peer-reviewed papers. The whole document went through four separate tiers of peer review, and received over 90,000 comments from 2500 separate reviewers. As geophysicist Paul Beggs asserts, this is 'arguably the most rigorous and transparent peer review process in the history of science' (Quoted in Feldman 2010).[18] This makes the IPCC a uniquely trustworthy source of information. We should take its reports as authoritative, and this means that facts about climate change as a *physical* phenomenon—albeit one with manifold ramifications in other domains—are primary.

But the trust we justifiably extend to this body's pronouncements need not be blind. We have seen that the IPCC makes errors. There is nothing fundamentally worrisome about this, however, so long as we become aware of the errors through the updated findings of peer-reviewed research. Indeed, this is the way science should work. What we should not accept are attacks on the science of the IPCC that have not gone through this process. Moreover, even if many of its findings are revised by subsequent research, the IPCC reports remain crucial documents. Many of the modifications, after all, are produced on the back of the research included in the IPCC reports. There is good reason to believe that absent the reports, these new findings either would not be produced at all or, if they were, that they would largely fall on deaf ears. This latter point is important: if nothing else, the IPCC is valuable for having brought the phenomenon of climate change to broad public attention (though I think its value goes beyond this).

In sum, if we want to know what the non-expert epistemic agent should believe about climate science, we should assert that she has sound reasons to trust the findings of the IPCC as modified by subsequent peer-reviewed research. In spite of the fact that the truth is in this case highly socially refracted we can tell a story that highlights the complex but guiding role of reason in the process of belief acquisition. The agent I have described, the one who successfully steers herself between the Scylla of full dependence and the Charybdis of full independence, has done something right and admirable. In describing her accomplishments we do not require the aid of extra-philosophical explanatory tools. However, there is an important distinction between beliefs whose etiology can be described this way and those whose production has been interfered with by non-rational forces. This distinction is the basis of Merton's A-Rationality Principle. As James Robert Brown understands the principle, 'If a rational explanation for a scientific belief is available, that explanation should be accepted; we should only turn to non-rational, socio-logical or psychological explanations when rational accounts are unavailable' (2001, 122). On this conception, those who believe in climate change (in the right way, to the right degree, etc.) have had their beliefs caused by the evidence, whereas the deniers' beliefs have been caused by one or more non-rational source, a fact which requires explanatory recourse to sociology, psychology (etc.) in these cases. All the forms of denial I have so far described involve the belief-formation process being hijacked by non-rational psychic forces. This is precisely why an appeal to epistemic vice can be so illuminating in this area. The same is true of self-deceptive denial, though we will see that this form is much more complex than the others.

5.4 Self-Deceptive Denial

It's hard to believe in anthropogenic climate change. One of the most fre-quent responses we encounter to the idea is that variability is a natural feature of the planet's biogeochemical rhythms and that this latest manifestation of weird weather is no exception. This thought is sometimes supported by the claim that it is simply beyond the scope or power of a single species to fundamentally alter these rhythms. The Earth is a colossus we humans are too puny to affect. The notion that humans are not only altering Earth's systems in measurable ways but that our interventions are a threat to the civilization we have built—the one that has provided us with the tools and methods for discovering how we are altering the Earth, for example—is, most often, dismissed as alarmism. I point to these mundane responses because they reveal something generally important about the way we believe, or refuse to believe. When it comes to testimonial knowledge, most of us assume that really big, important truths will have been revealed by the sources we conventionally trust. If someone then informs us of a really big, important truth not covered by one or more of these sources—or denied outright by them—we declare the proposition incredible and move on.

Goldberg has analysed this phenomenon under the rubric of 'coverage-supported belief' (CSB). Most of the philosophical literature on testimony is focused on the conditions under which a certain testimonial source should be believed. Goldberg, by contrast, asks why in some cases we refuse to believe. He notes that our reasons for refusing to believe often point to the fact that none of our sources have testified that-p, but that they would have done so if it were the case that-p. So it must be the case that not-p (Goldberg 2011, 95). According to Goldberg, there is nothing in principle wrong with this sort of structure but there are five conditions that must be met in order to justify the key claim—that my trusted sources would have testified that-p if it were the case that-p—in any particular case. The source in question must exist, it must be timely and reliable in uncovering the facts, it must make the facts plainly available, the facts must not have been revealed already, and there must have been sufficient time for the facts to have been revealed (Goldberg 2011, 97–9).

As I have just hinted, something like a CSB structure seems to be at work with climate change. Perhaps the relevant thought is something like, 'if things were as bad as the alarmists say, then there would surely be unanimous agreement among all relevant experts—scientists and governments, but also the media—about the scope of the threat and how it should be met'. Absent this level of agreement, one concludes that the threat is not real or is overblown. Now, it is relatively easy to show that the five conditions are not met in the case of climate change. Oreskes and Conway (2010) have demonstrated convincingly that the media, for example, has badly misinterpreted both the level of agreement among climate scientists about the fundamentals of climate science and the science itself. And many governments have clearly been willing to ignore climate science when recognizing it has been thought to conflict with various *raisons d'état*. The unreliability of these sources on this issue is blindingly obvious to anyone willing to look. The relevant CSBs are therefore clearly not justified on Goldberg's terms.

And yet, the relevant information about the threats of climate change *is* available in the form of IPCC reports and the subsequent corrections made to them, as we have seen. Moreover, much of this information has been ably popularized. So we need to go beyond purely epistemological analysis in order to find out what has gone wrong in our belief-formation processes. This follows from the A-Rationality principle: when reason has been interfered with, we need the explanatory help of (at least) sociology and psychology. So in what remains of this chapter, I'm going to lean on some recent work on the phenomenon of climate change denial from these two disciplines (sociology in this section, psychology in the next). Part of the problem, of course, is precisely the refraction of information we have been looking at. Many conventionally authoritative sources are caught up in a wider web of denial. But the more important point to make is that our lifestyles are so densely permeated by the use of fossil fuels that even those of us who are seriously concerned about climate change don't do anything meaningful about combatting it. How should we understand this?

In her ethnography of the people of Bygdaby—a ski village in Norway—sociologist Kari Marie Norgaard describes people who are worried about climate change but also know that their prosperity rests on the exploitation of Norway's substantial oil resources. These are for the most part educated and progressive people, so they are a window on the global prosperous as a whole. In her analysis of them Norgaard reveals 'the process through which climate change is kept out of the sphere of everyday life' (2011, 123). The main task, as she sees it, has to do with emotion management. Climate change is very real for these people because warming weather has wreaked havoc on snow conditions. Skiing is a way of life for the Bygdabyingar, so they need to engage in some pretty tortured emotional and cognitive gymnastics in order to keep bad thoughts at bay. Through a series of interviews with the locals, Norgaard shows us the ubiquity of irony, cynicism, and humour as modes for deflecting uncomfortable thoughts. Citizens frequently invoke the need 'not to get depressed' by thinking too much about these matters, they turn their progressive energies to unrelated matters (working for human rights in other countries, for example), actively immerse themselves in the past traditions and myths of their community (so as not to think about the future), focus intently on the local rather than the global, and more.

The climate change denial of the citizens of Bygdaby is qualitatively unlike that of the other agents of denial we have examined. Here is Norgaard's description of the people she has studied:

[H]olding information at a distance is actually an active strategy as they negotiate their relationships with climate change. The notion of socially organized denial emphasizes that ignoring is done in response to social circumstances and carried out through a process of social interaction ... That is to say, at this time, people had a variety of methods available for normalizing or minimizing disturbing information. These methods can be called strategies of denial. It is significant that what I describe as 'climate denial' felt to people in Bygdaby (and, indeed, to people around the world) like 'everyday life'. Nonresponse to climate change was produced through cultural practices of everyday life. (2011, 121)

What Norgaard shows us corresponds to a certain understanding of the mechanism of self-deception. To explain this, and show that it is a plausible way to frame one face of climate change denial, we need to delve briefly into the theory of self-deception. Parallel to my analysis of the mechanics of moral weakness in chapter 4, there are two broad ways of thinking about self-deception: the agency view and the anti-agency view. According to the agency view, there is, in Pears' phrase, 'a separate center of agency within the whole person [that is] from its own point of view entirely rational' (1984, 87). This center of agency 'wants' the main system to form an irrational belief, so it suppresses the intervention into the cognitive process of cautionary beliefs which would have the effect of subverting this goal. In other words, the person, or some sub-agency within the person, brings it about that she holds a particular belief. Since it demands that the self-deceived agent has a clear-eyed view of the truth of a proposition whose truth she *also* denies, this view is deeply puzzling.[19]

For the same sorts of reasons that it is difficult to accept strict or open-eyed weakness, we probably have good reason to reject the agency view of self-deception. Fortunately, Mele has given us a compelling anti-agency theory of self-deception. He lists four conditions jointly sufficient for my being self-deceived: (1) the belief that-p I acquire is false; (2) I treat data relevant to the truth of p in a motivationally biased way; (3) this treatment is a non-deviant, i.e. non-accidental cause of my acquiring the belief that-p; and (4) the body of data I possess at the time provides greater warrant for not-p than p. This list is focused on the problematic treatment of data. When we treat data in a motivationally biased way we may become self-deceived about what they report. No reference to an intention or strategy to create or sustain the false belief is required (Mele 2001, 26).[20] On the anti-agency view we need only say that the self-deceived have desires (or other non-rational psychological states) that cause them to skew evidence and therefore to develop false beliefs.

The anti-agency view of self-deception is a powerful theoretical tool for understanding how so many of us have mishandled the data presented to us by climate scientists. There are many ways data can be mishandled, but I want to focus on one general way this happens, one that strikes me as particularly illuminating for climate change and the global prosperous. According to Annette Barnes, self-deceptive beliefs are the causal product of anxious desires, and their function is to mitigate the psychological impact of those desires. This is how Barnes characterizes the view:

> In self-deception, the input, so to speak—an anxious desire that-not-q—always results in an output belief that-p. This output belief, if it is to be a self-deceptive one, must function to reduce the anxiety that produces it. The resultant self-deceptive belief, however, is arrived at because the person's belief-acquisition process is skewed, because a bias in favor of beliefs that reduce anxiety is in place. Perception, memory, imagination, reasoning, etc. can all be distorted by this bias. (1997, 79)

Since the specific ways agents respond to anxiety about climate change will differ markedly from one cultural context to another—that is, the relevant distracting activities will vary widely—if we are to grasp this problem we really do need much more of the sort of thick sociological analysis Norgaard provides. This is well beyond my ken, so to provide some symmetry with Marc from Chapter 4, I offer instead another character sketch.

Imagine Janet, an American, who has a sense that the threats from climate change are very real and have the potential to bring catastrophe, if not to her corner of the world then at least to large swathes of the developing world. Being a decent person, she has an anxious desire that this not happen but finds the problem too large and disturbing to come fully to grips with. The anxious agent often feels as though she has lost control over what she takes to be an important part of reality. I think that many of the strategies of denial discussed by Norgaard represent an attempt to regain control of things. As I have been emphasizing, it is important that so many of these activities are social in nature: there are powerfully affirmative forces of group psychology at work here. The problem is that the control is asserted with respect to concerns or issues other than the one causing the anxiety, but the feeling of satisfaction caused by the assertion of control in these areas is then, as it were, projected back onto the anxiety-causing concern. The agent now feels as though that phenomenon too is under control. When things are in fact bad in the primary area of concern, this new feeling can result in self-deceptive belief formation.

This is what happens with Janet, let's suppose. She finds like-minded people—in her church, her political party, her family, her law firm, etc.— and together they talk about and perhaps even solve a number of issues and problems unrelated to climate change. Buoyed by this social support, and these positive results, she has created psychological space in herself for beliefs about climate change that are more optimistic than the evidence warrants. She has pushed her original anxiety aside for a time. But reality has a way of seeping back in. Arctic ice keeps disappearing, temperature records are continually being shattered the world over, giant typhoons sweep across the South Pacific Ocean (and so on), all of which keeps the embers of anxiety burning in Janet and perpetuates a bias in her toward further distracting activities and the self-deceptive beliefs they enable. It is important to emphasize that the sort of distraction at work here does not result in total non-confrontation with the issue. Rather, it leads agents to develop less threatening beliefs about the issue than the truth warrants. So in a psychological state like this how, specifically, might Janet think about climate change?

Since I am trying to mark her off from the kinds of climate change deniers examined in Section 5.2, let's suppose that she is, generally speaking, a truthful person and is concerned about climate change in particular. Because it is so ubiquitous I also want her to endorse the promise of decoupling, examined in the abstract in Chapter 4. So she might affirm these propositions:

(P1) Anthropogenic climate change is real, and is a serious threat, but it is such a large and complex problem that nobody can properly be blamed for it.

(P2) It is important for us to break the link between our consumption and our emissions, perhaps by developing our renewable energy resources more aggressively. There is no reason to believe this will put downward pressure on our lifestyles.

It is no accident that (P1) is the darling proposition of many American politicians.[21] It represents a dangerous development in our thinking about climate change because, at least as it is currently invoked, it essentially erases *both* blame *and* historical responsibility for the problem. As Shue has argued, it is important to attend to the distinction between blame and responsibility in this area: whereas it may be the case that Americans cannot be blamed for climate change it does not follow that they are not disproportionately responsible for it. Ascribing responsibility is a matter of specifying an agent's causal or historical role in bringing about a state of affairs and does not as such entail that the agent merits (say) punishment for playing that role, though it does entail that the agent must bear some reparative costs.[22] We should not be surprised to see (P1) arise most prominently in the country bearing the greatest historical responsibility for climate change. In asserting it, politicians are trying to appease voters who understand that climate change is a serious problem but who also refuse to make significant sacrifices to their current way of life in order to address it. What better strategy than to suggest that Americans like Janet are not to blame for climate change? In this case, they are not morally required to incur significant economic costs in order to help ameliorate the material conditions of either the current global poor or future generations.

As I have said in setting up Janet's beliefs this way, these considerations bring us back to a key theme of Chapter 4 because (P1) and (P2), taken together, constitute a thinly veiled assertion of the priority of what Broome calls 'efficiency without sacrifice'. That is, while (P1) provides cover for the abdication of historical responsibility for climate change, (P2) expresses the idea that it would nevertheless be prudent to bring greater efficiency to our exchanges. Happily, this can be done on the cheap. It is interesting that some of those American politicians currently refusing to 'play the blame game' are in favour of otherwise progressive energy policies, like moving from coal to renewables.[23] But even in the unlikely event that the widespread adoption of these propositions helped bring about the emergence of a more robust renewables market in the U.S., the refusal to fund green development in poor countries while U.S. consumption continues to embed emissions from those countries would allow climate change to worsen.

Recall that the anti-agency view of self-deception requires, among other things, that the relevant belief is false and is contradicted by the available evidence.[24] Does it make sense to say that (P1) and (P2) might be true or false in this sense? After all, though each contains both descriptive and normative elements, their main force is clearly normative or prescriptive, and philosophers have always struggled with establishing truth conditions for such claims. To say that both propositions are 'false', however, does not need to be supported by the strong claim that moral truths are part of the furniture of the world, if by this we mean that they transcend all historically particular human communities. Instead, we can point to the *unreasonableness* of agents like Janet rejecting the polluter pays principle (PPP) given their other beliefs and values.[25] According to the PPP, there is an intimate connection between an agent's (or group's) historical responsibility for a given undesirable state of affairs and that agent's (or group's) moral obligation to bear a disproportionate share of costs for repairing that state of affairs (other things equal). The PPP is clearly one of our considered moral judgements and we should not be shy about talking of such judgements as being capable of contradicting some of our other beliefs or practices. It is difficult to credit the reasonableness of any outright rejection of the notion that you are responsible for fixing the things you break.[26] With this notion in hand we can see exactly what is wrong with (P1) and (P2).

If it is a proxy for the denial of responsibility, as I've suggested it currently is in practice, (P1) is flawed because Americans *are* disproportionately historically responsible for the broken climate. Between 1850 and 2002, the U.S. produced 29.3 per cent of the stock of greenhouse gases, an amount that dwarfs the historical contribution of any other single country (Herzog et al. 2005, 32). Nor is the responsibility cancelled because Americans did not intend to cause climate change or were ignorant that they were doing so. Agents can justifiably be held responsible for harms they non-intentionally cause, like the driver who causes an accident and has to pay damages. The ignorance claim is no more plausible because at least since James Hansen's congressional testimony in 1988 we should have been fully aware of what we are doing. If the further claim is made that it is not right to hold a present individual responsible for the past wrongs of the collective of which he is a part, we should remember one of Socrates' reasons for refusing his friends' invitation to escape after he had been condemned to death by the Athenians. In the *Crito* Socrates argues from his jail cell that one cannot justifiably accept the material benefits of one's collective while refusing its moral and material burdens, even if these involve heavy costs (one's life, in the extreme).[27] From these points it follows that the U.S.—and by extension ordinary citizens like Janet—should bear a disproportionate share of the costs for climate clean-up. And contra (P2), as I have argued in Chapter 4, these costs go well beyond implementing efficiency measures in the domestic U.S. energy sector. They cannot be fully met without a significant reduction in consumption among Americans.

Everything just described is within Janet's cognitive reach. She's not an unintelligent person so in some sense she gets all of this—even if she's never read Plato—but she also perceives dimly that the only real solution to the problem will involve significant sacrifice on her part. And so she becomes palpably troubled, as a result of which she flees the facts and loses herself in socially supported distracting activities. *This* move then allows the less threatening beliefs—(P1) and (P2) in this case—to push out the more threatening ones. If this process sounds mysterious, it shouldn't. We all engage in it to some degree. Think of the husband who has discovered disturbing evidence of his partner's infidelity. Or rather, he discovers *clues* that might lead him in this direction but refrains either from fitting all the clues together or tracing some of them to their logical conclusions. Propelled by anxiety, he might instead seek out the comfort of an old mate, one skilled at seeing the sunny side of things. This friend takes him out for dinner and a bottle of good wine, stokes his nostalgia by reminding him of the rich and trusting history he and his partner share, and so on. Through a carefully coordinated process of distraction and strategic manipulation of emotion he helps the poor man convince himself that what he has discovered about his partner is not what it seems.

This analysis does not apply to just anyone unable to appreciate the full force of the problem. Two distinctions are important here. The first is between the self-deceived and the merely ignorant. The key factor in this comparison is that in the former case the agents' commitment to certain evidentiary norms gives real bite to the notion that she possesses the relevant data, the ones contradicting her self-deceptive belief. In turning away from that evidence, the self-deceived agent is neglecting to attend to the sorts of reasons she usually takes seriously. Think of this in relation to Janet's way of understanding the PPP. If she or her child breaks something in a store she expects to have to pay for it, or if a copper mine's dams are breached, spilling toxic sludge into adjacent waterways, she expects the mine's owners to pay for the clean-up. Moreover, in the latter case, although she realizes that the costs will be shifted to shareholders who were not directly responsible for the disaster, she judges this irrelevant to the question of responsibility. I find it incomprehensible to suppose that Janet cannot sense that these principles commit her to a certain way of understanding her group's complicity for the damaged climate. This is why it makes psychological sense to suppose that she becomes anxious on encountering the relevant facts.

The problem of anxiety brings us to the second distinction, which is easier to deal with. There's a crucial difference between the self-deceived and the three other kinds of climate deniers I have examined. With these deniers we do not need to point to anxiety in the etiology of the false beliefs. There is skewing of data in all four cases, but the psychological forces responsible for this are importantly different in the first three on the one hand and the self-deceived on the other. Indeed, one of the problems with the first three types of deniers is that they are *insufficiently* anxious about the threats posed by anthropogenic climate change.[28] But fictional agents like Janet aside, we might wonder if any of us is suitably anxious about the problem. As it happens, though the phenomenon has not been subjected to rigorous study, anecdotal evidence among mental-health care professionals is emerging that more and more of us—especially young adults—*are* feeling increasingly anxious about climate change (MacDonald 2014). This anxiety is not just about perceived threats to one's own livelihood and security, but also involves a dim awareness of partial responsibility for the problem. Support for the anecdotal evidence can be found in a global PEW Research survey conducted in the spring of 2013 asking people in thirty-nine countries about their perception of 'major threats'. In nearly every region of the world apart from the U.S. and the Middle East, climate change topped the list, ahead of international financial instability (this only a few years after the global economic crisis), Islamic extremism, Iran's nuclear programme, and other options (Wike 2014).

Let me make two quick points about these data. First, even in the U.S. and Middle East a very large number of people do see climate change as a major threat (40 per cent in the U.S., 42 per cent in the Middle East), even if it is not believed to be the biggest one. Second, in spite of the American and Middle Eastern numbers, a median of 54 per cent of the surveyed global population thinks that climate change is the major threat we face. That's a very large number. So why have we failed to act? Although the manner in which fossil-fuel interests have distorted our political culture is a large part of the problem here, it is not the whole story. The question I have about that portion of the 54 per cent that constitutes the global prosperous (which I'm not claiming is the whole 54 per cent) is how many of them are willing to radically alter their lifestyles in order to do something meaningful about climate change? Again, it's impossible to say with precision, but since the problem *is* getting worse, I think the answer is, clearly, not enough of them. But that answer invites an analysis like mine because it means that the anxiety revealed in the survey—and what is a threat-perception if not an expression of anxiety?—is very likely affecting our response to the issue in significant ways. Specifically, our anxiety is, I suggest, causing us to turn away from what we should be seeing clearly and acting aggressively to address.

5.5 Anxiety, Negation, and Disavowal

Though some cases of self-deceptive belief-formation can be trivial, it is clearly possible for the biases that Barnes and Mele have identified to have extremely negative consequences. They might, for instance, interfere with one's belief-forming processes generally, or undermine one's integrity or agency, or they might constitute an abdication of responsibility (Barnes 1997, 170). Because of these possibilities, self-deception is often a blameworthy state for an agent to be in. That is a familiar enough thought. But let's set it aside for a moment and see if we cannot locate some purchase for optimism here. We should remind ourselves that split selves, unlike wicked or malevolent selves, can be repaired. The self-deceiver has not *abdicated* truthfulness. This is why Williams says that 'self-deception, which is one thing that the accurate agent must avoid, is an homage that fantasy pays to the sense of reality' (2002, 135). The best place to explore this idea is the psychoanalytic literature on climate change.

If we are anxious about anthropogenic climate change in the sense required for this analysis it must be because we perceive, however dimly, that something of very deep value has been damaged by our activities. Those working on the problem of climate change in the psychoanalytic tradition are important because they see better than anyone else what is involved in this phenomenon. The focus of much of this analysis is on the way selves have become split in the face of this reality, because it is simply too difficult to bear the full emotional weight of the loss. Two themes stand out. The first is the novel interpretation these researchers and practitioners provide of apathy or inertia about climate change, the second the idea that resolution of the problem involves embracing, rather than fleeing, the pain and anxiety involved in the contemplation of it.

When they discuss the issue at all, most philosophers tend to understand our apathy or inertia in highly moralized terms.[29] This is understandable to the extent that we are concerned with the negative outcomes of such psychological states. But things look different if we look more closely at those psychological states themselves:

> The operation of . . . 'turning a blind eye', which describes the not-noticing that allows for the failure to give meaning to known facts and to the disavowal of responsibility is also pertinent. This type of defence comes into operation not from moral failure, but from the intolerable nature of the anxiety that has to be faced. It generates too great a degree of insecurity and persecution and places too heavy a burden of guilt. (Rustin 2013, 107)

Apathy is one upshot of this psychic retreat. Agents experiencing anxiety consequent on the perception of damage they do to a good (for which damage they consider themselves at least partly responsible) will, as we have seen with Janet, engage in myriad forms of turning away from the perception. Apathy is a broad phenomenon because it can manifest not only in paralysis but also, more interestingly, in frenzied activity. As long as the latter can distract the agent from her anxiety it serves its function. If this is correct, then unintelligent green consumerism—the sort that's easy prey for greenwashing advertisers—can be as apathetic as is simply throwing one's hands in the air. As such, apathy is generally treated by psychoanalysts not as evidence that the agent cares too little about an issue but, on the contrary, that she cares too much about it.

This brings us to the second broad theme in the psychoanalytic literature. Following Melanie Klein, Sally Weintrobe has argued that there are two forms denial can take: negation and disavowal. The first is healthy, the second pathological. Negation is merely 'our first response to reality when it faces us with shocking losses and changes' (Weintrobe 2013, 38–9). Negation is thus self-protective, and to the extent that it results in self-deception, so too is this reaction. Part of the process of mourning in its ideal form is that we are supported emotionally by significant others in our lives, whose job it is to help us process change. This can go beyond intimates and extend to social groups and even political leaders. The key is that the denial in negation is a short-term coping strategy that is meant to fade away so that reality can again be faced.

But as far as climate change denial is concerned there is no guarantee that we won't opt instead for a more radical form of splitting: disavowal. This form, as Weintrobe argues, is 'part of a pathological organization' (2013, 39). It is a radical, and radically narcissistic, retreat from reality. Here, some of the same distractions as were employed with negation might still be used, but the quality of falseness in the belief-set might also result in more distorted beliefs. For example, denial of responsibility and cynicism might come to the fore more aggressively than before. The potential result is twofold: a short-term quelling of anxiety that is, initially at least, more effective than that involved with negation; and a more arrogant approach to reality that stresses omnipotent-thinking. If Janet's negation becomes disavowal, she might resort to one or more of the other forms of denial we have looked at here. In this vein, we should, I suggest, be especially attentive to the manner in which disavowal can work hand in glove with official bullshit to unfit us for the truth.

The problem with disavowing responses is that, considered as measures designed to silence anxiety, they won't last:

> Disavowal is a poor means of lowering anxiety in the long term. Because disavowal does nothing to address the real causes of anxiety, it can lead to an escalation of underlying anxiety that can feel increasingly unmanageable. The more disavowal is allowed to proceed unchecked by reality, the more anxiety it breeds and the greater the danger that the anxiety will be defended against by further defensive arrogance and further disavowal. Disavowal leads to a vicious spiral . . . (Weintrobe 2013, 39)

The key to avoiding this outcome is to catch denial at the stage of negation, and move it back to avowal and reparative activities before it degenerates into disavowal. But this can happen only if the pain of loss is admitted and embraced. This is why Rosemary Randall says that in order to deal with what she calls 'ecological debt' we must actually face our sense of loss, trauma, despair, guilt, and rage, and indeed that these emotions should be 'extended to the rest of the population' instead of our focusing on 'upbeat solutions and easy steps' (2013, 100). James Hansen has made the provocative suggestion that with respect to the climate we are somewhere between a tipping point and a point of no return (Hansen et al. 2008). What he meant by this is that we have tripped some positive feedbacks in the climate system—Earth's albedo, for example—but that we are not yet faced with runaway climate change, where no matter what we do warming will accelerate. For self-deceptive climate change deniers the psychological analogue of this precarious geophysical position might be something like the space between negation and disavowal. Many of us walk this psychological knife-edge.

If self-deception is produced by a bias toward anxiety-reducing beliefs, then it is important to look for solutions to it in the sphere of character alterations. A bias, after all, is a disposition with characteristic expressions in thought, emotion, and behaviour. What virtues might we seek to cultivate to help prevent the slide to disavowal? A clue is provided by Weintrobe's reference to the arrogance involved in disavowal. The antidote to arrogance is humility, which once again reveals itself as a key intellectual virtue. It works in this context by helping the agent cultivate a 'sense of reality'. Bernard Williams puts these points in terms of the relation between the will and the world:

> What the inquirer must have in mind, as a condition of getting himself and his beliefs in the right relation to the world and to his will, can be summed up in these

terms: some things in the world he can affect and most others he cannot; with regard to what he cannot affect and knows he cannot affect, a want can only be a wish, and a belief cannot properly be dependent on a wish. (2002, 135)

Curbing the demands of narcissism and the tendency to omnipotent-thinking involves learning to observe this distinction carefully. But facing the painful emotions consequent on our recognition of the damage we are in the process of doing also demands extraordinary courage. These, I suggest, are the two virtues most required to help block the slide to denial as disavowal. Truthfulness about climate change cannot do without them.

The most important general insight from the psychoanalytic literature on climate change is that it is probably futile to try to combat self-deception directly. A better strategy is to work on the emotions sustaining it. *Facing the anxiety we feel about climate change means looking at the problem—including our own role in producing and worsening it—squarely.* The more firmly we stand in the light of our anxiety and ambivalence, the less scope we give to self-deceptive belief-formation. There's a chance that the stance I have recommended could induce panic or paralysis, but the virtue of courage should also be a check on that. In any case, the evasive techniques involved in disavowal are surely a more dangerous kind of paralysis, since they masquerade as meaningful activity and engagement.

5.6 Conclusion

I have not offered a general account of the intellectual virtues or vices, putting forward instead an applied regulative epistemology for the climate crisis. We need to do a better job instantiating the virtue of truthfulness most of us already claim to endorse. The argument has been that the fullest explanation of our epistemic failing should make reference to the intellectual vices. This is because the language and concepts of traditional epistemology—truth, warrant, justification—simply do not cut deeply enough on matters like this. They do not account adequately for the 'added value' of superlative epistemic achievement or, more to the point, the 'subtracted value' of egregious epistemic failure. And the way we have for the most part structured our beliefs about climate change is most certainly an example of egregious epistemic failure, whatever else it is.

These points apply most obviously to the first three faces of denial, but we also need to highlight the role of self-deception in helping to perpetuate the epistemic status quo because it too involves a failure to live up to the ideal of truthfulness. Moreover, the account of self-deception presented here is of a piece with the account of moral weakness in the previous chapter. Both the attention-withdrawal model of moral weakness and the anti-agency view of self-deception assume that agents go astray by *avoiding* a stark confrontation between the relevant psychic elements: conviction and choice on the one hand, graspable evidence and belief on the other. This is what makes it possible to say that in the case we are examining it is one and the same agent who is both weak and self-deceived. The paradoxical conclusion at which we have arrived, however, is that our best hope for real change may rest precisely in these facts about us, for the weak are not yet wicked and the self-deceived have not yet given themselves over to fantasy. That we are still moored in this way to the good and the true is some cause for hope, a separate virtue to whose analysis we now turn.

6

Hope

6.1 Introduction

In *The Revenge of Gaia*, Lovelock provides a memorable image of what the future might hold for us in a world severely blighted by climate change. In the scenario he describes the human population has been pushed to the high Northern latitudes:

> Meanwhile in the hot arid world survivors gather for the journey to the new Arctic centres of civilization; I see them in the desert as the dawn breaks and the sun throws its piercing gaze across the horizon at the camp. The cool fresh night air lingers for a while and then, like smoke, dissipates as the heat takes charge. Their camel wakes, blinks and slowly rises on her haunches. The few remaining members of the tribe mount. She belches, and sets off on the long unbearably hot journey to the next oasis. (2006, 159)

Here, we can imagine that an archipelago of circumpolar cities—Indiga, Pesha, Point Hope, Wainright, Iqaluit, etc.—have become the new 'centres of civilization'. Lovelock's portrait of the future is arresting, but tells us nothing about how to evaluate that future world morally. In spite of the reference to a new 'civilization' we cannot say if this a vision of hope or of despair. Because it is important to imagine the *moral* world we are bringing into being, complete with details about who gets what share of scarce resources and why, in the course of this chapter we will need to supplement Lovelock's polar tableau with other ways of imagining the future.

In Chapter 2, I argued that, as bad as things are, we ought not to indulge practical pessimism about climate change. That claim entails that some more positive attitude towards the future we are creating is justified, but leaves open the question what precise shape this attitude should take. Optimism, based on the belief that things will probably turn out well for us, is, alas, surely unjustified. Everything we now know about the severity of the climate crisis suggests that the probabilities simply do not come out this way. Still, even the direst popular accounts of the science of climate change contain at least implicit expressions of hope that we can avoid total catastrophe (however this is understood).[1] Expressions of hope also abound among climate ethicists. For example, Stephen Gardiner's recent work is unsparingly critical of our 'moral corruption' on this issue and yet he too believes that his diagnosis provides hope for a solution to the problem (Gardiner 2011, 184). All of this highlights the need for a philosophical assessment of the grounds for hope in the face of the climate challenge. And yet, though philosophers have for some time—if only sporadically—been reflecting on the nature of hope,[2] a philosophy of hope for the climate crisis is almost entirely absent from the philosophical literature.[3]

The following analysis allows us to identify the complex object our hope should take. It is that we will (a) decarbonize our economy quickly enough to (b) preserve a meaningful normative continuity between our generation and subsequent generations, where the connecting thread is the impartial or cosmopolitan sense of justice. However, if we fail utterly in respect of (a) the prospect of our achieving (b) is doubtful. The chapter is structured around a description of two possible futures. The first is a moral, if not a literal, doomsday scenario. Here, the people of the future have lost normative contact with us because in their world social benefits and burdens are distributed entirely on the basis of chance or official caprice. This outcome—the disappearance of justice as both a practice and a normative ideal—is self-evidently dreadful for those living with it. However, drawing on recent work by Samuel Scheffler, I show that many of our own projects and goals would also be drained of meaning were we to come to believe in the likelihood of this outcome.

The second possibility is where future societies have been forced to allocate resources on the basis of triage and a survival lottery. This is a better outcome than the first one because, as Catriona McKinnon has shown, the survival lottery can express a commitment to impartiality. But it is still bad insofar as the lottery is run on the back of a system of triage, which is not just. Because the second outcome could easily degenerate into the first—indeed they could co-exist in different parts of the world—my argument is that we should focus our hopeful energies only on the task of decarbonization in an effort to avoid both. But it may not be possible to achieve this unless we are willing to alter almost everything about the way we now live. Our hope therefore must be 'radical' (in Jonathan Lear's sense) which, I show, is equivalent to saying that it must be revolutionary. I begin with a general analysis of the virtue of hope.

6.2 The Virtue of Hope

What is hope? Suppose I had been hopeful that the climate negotiations at the Copenhagen Climate Conference in 2009 would bear fruit in the form of a robust and binding successor treaty to the Kyoto Protocol, complete with meaningful emissions reduction targets for developed and developing countries, agreements to assist developing countries in their adaptation struggles, an agreement to end deforestation in the tropics, and so on. Let's linger on this example and examine the account provided recently by Australian climatologist Tim Flannery about his own shifting attitudes toward the conference. I begin with two quick methodological points. First, I am imaginatively reconstructing Flannery's lengthy anecdote as contained in his recent book *Here on Earth*. My point is not to suggest this is exactly how Flannery himself would see things, only that it is a plausible interpretation of what he does say. Second, and more importantly, the suggestion is not that such a progression is inevitable. There are many possible permutations of it. For example, confidence or optimism might get hijacked by wishful thinking, which would change everything. Or demoralization about an outcome might give way to despair rather than renewed hope. My primary purpose here is neither to play the amateur biographer nor to lay down laws of psychic development but to begin the task of elucidating the nature of hope through contrast with its cousin-concepts.

Flannery published the enormously successful *The Weather Makers* in 2007, and then stopped doing science to focus on mobilizing progressive corporate leaders to help tackle climate change. September 2009 was a high-water mark for Flannery. At this point, Flannery says that he had 'never felt so optimistic about the prospects of humanity overcoming its greatest challenge' (2011, 247). This is clearly not hope but optimism. A few weeks hence, however, this optimism was destroyed when people began to realize that the U.S. Senate would not pass its climate legislation in time for the meeting. As Flannery puts it, 'it felt as if the world was being held to ransom by a few holdouts whose reflex beliefs in the "survival of the fittest" permitted violations of the common good' (2011, 247). But confidence that the outcome will prevail is not thereby abandoned because optimism is replaced by hope. There are two things to say about this development. First, it reveals that we were correct to think of the first stage as involving optimism rather than wishful thinking because the latter attitude is not sensitive to evidence-based falsification the way the former is. Second, it reveals two important things about hope: (a) that it is based on less secure evidentiary grounds than is optimism; and (b) that with it the agent nevertheless maintains a strong commitment to the desired outcome.

According to the 'standard account' of hope, A hopes that p if and only if (1) A desires that p; and (2) A believes that p has some degree of probability, however low.[4] Criterion (1) is an obvious requirement since on any account of hope it is, in part, a species of desire. If one did not have a pro-attitude to an outcome it is hard to imagine what it might mean to hope for it. The reference in criterion (2) to belief in an outcome's probability is meant to distinguish hope from standard accounts of justified belief where justification works to inspire a believer's confidence in the state of affairs picked out by the belief. That is, if the probability is very low, the non-hopeful believer considers the outcome simply unlikely whereas the hopeful believer as such places greater confidence in it. How is it that hope, thus construed, differs from self-deception or wishful thinking? The answer is that the self-deceived agent and the wishful thinker violate a norm of theoretical reason by mis-characterizing relevant probabilities.[5] The belief component of hope is reason-responsive and sensitive to evidence. The hopeful cancer patient does not believe, falsely, that her chances of beating the disease are greater than they in fact are, but that she will survive even in the teeth of poor odds.

We should add that even where belief in an outcome falls short of certainty (or knowledge) but still inspires a high degree of confidence, we rarely express hope for it. I may not know that the contractor I have just hired will not swindle me, but since he has worked for me in the past and has never yet done so I am confident he will play me straight this time too. I don't hope for this, I simply expect it. Still, the important point is that although one can judge an outcome extremely unlikely or doubtful, it would not for all that be unreasonable to hope for it. While neither the belief nor the desire components of the standard account are unique to hope, the psychological state described in their combination is *sui generis*. Further, the psychological literature on hope demonstrates that this peculiar mental state tends to produce in its bearer a high degree of perseverance vis-à-vis the object of hope. That is, those with hope 'generate more effective routes' to their goals than ordinary desirers, 'especially under impeding circumstances' (Snyder and Rand 2003, 821).[6]

However, the standard account does not appear to provide a way of distinguishing between hope and despair. If Bob and Sue both have a very strong desire for outcome x, and both have the same understanding of the relevant probabilities, how do we explain the fact that, say, Bob is hopeful about x while Sue, though she desires it, also despairs of it? The standard account obviously cannot make sense of this difference because by hypothesis both agents meet requirements (1) and (2) above.[7] The missing element, according to Martin, is a 'way of seeing one's situation' that is typical of the hopeful person: '[T]he hopeful person takes a 'licensing' stance toward the probability she assigns the hoped-for outcome—she sees that probability as licensing her to treat her desire for that outcome and the outcome's desirable features as reasons to engage in [specific] forms of planning, thought and feeling' (2014, 35). This is the 'incorporation thesis', according to which the hopeful agent 'incorporates' her hopes into her psycho-behavioural economy: her emotions, beliefs, actions, etc. So far, this does not appear to ground the distinction between Bob and Sue. Can't Sue say that she has also incorporated her despairing attitude into her plans, thoughts, and feelings?

The best way to distinguish Bob and Sue is to say that Bob's hope can be sustained only insofar as he draws into his psycho-behavioural economy reasons that are extrinsic to the specific belief–desire pair that constitutes the hope. That is, he must come to believe, independently of his desire for the outcome and his assessment of its probability, that the hope is instrumental to the promotion of some other end. Sue, by contrast, cannot generate hope precisely because her attention is riveted on the relevant belief–desire pair. She is so overwhelmed by the harsh conflict between what she wants and what the world is offering her that no extrinsic considerations can enter the picture. This distinction fits our intuitive understanding of such agents. The hopeful person seems to be able to abstract from his worries while the despairing person cannot tear her mind away from them. And while this explains why one agent can generate 'hope against hope'—that is, hope directed at very important matters in the face of especially bad odds (Martin 2014, 5)—while another cannot it also works in less dramatic cases. Wherever the probabilities invite pessimism, appeal to extrinsic considerations can bolster hope. It's just that the more dire the probabilities, the more difficult this is to accomplish. Hope against hope is an *exceedingly* difficult psychological state to sustain, which means that the extrinsic considerations supporting it must be especially powerful or compelling to the agent. This is important to note since our hope for a meaningful solution to the climate crisis must, I think, take this form.

Bob's ability to hope—even hope against hope—does not yet make him virtuously hopeful. There are two ways in which he can fail to be so. First, he can hope non-dispositionally. This would happen if the extrinsic ends which sustained the hope on one occasion failed to do so reliably. In that case we would be tempted to say that the hope was deviantly caused in Bob; that it is more the product of Bob's circumstances than his practical identity. Indeed, reliability thus construed strikes me as a good way to parse Martin's incorporation thesis. It is because the hope is taken up comprehensively by the agent that he is a reliable hoper. Knowledge of this fact about him will likely have implications for how others deal with him as well as how he deals with himself (affecting his trustworthiness or self-trustworthiness, for example). Second, even if he were dispositionally hopeful, his hope might be unethical. The ends he reliably summons to bolster his hope might be morally neutral or immoral, whereas virtues are morally praiseworthy dispositions. In this sense, hope is more like courage than justice: both hope and courage can be used for immoral or morally neutral ends, though justice is intrinsically praiseworthy.

To get a handle on how the disposition becomes a virtue, think of the wider circle of social relations in which Bob and Sue might move. Suppose that Sue's despair causes her to treat other people—her children, for instance—unjustly. She might decide that because the future is so bleak for her, nobody else should be happy either and so she spends her children's inheritance recklessly. Bob, on the other hand, might justify his hope by saying that it fortifies him to do the things he believes he ought, on independent moral grounds, to do. For instance, his hope for a better future might, as Calhoun puts it, 'second his commitment' to fighting poverty in his community (Quoted in Martin 2014, 85). And we should add that the moral reasons supporting hope might also be primarily self-regarding. Bob might think that proper *self*-respect requires hope. In this case, since they allow the agent to abstract from the relevant belief–desire pair in the right way, the demands of self-respect constitute the extrinsic consideration I have been talking about. Whether other-directed or self-directed, this turning of the mind's eye away from the problematic belief–desire pair and towards independent moral considerations is a key component of hope as a virtue. To summarize: A's hope, H, is virtuous if and only if (a) H is reliably incorporated into A's psycho-behavioural economy; and (b) H is supported in A by specifically moral reasons.

Let's return briefly to Flannery to complete our inventory of moral psychological attitudes. As he tells it, 'the final blow' to his hopes came with the realization that political machinations were going to scuttle any chance for a meaningful deal in Copenhagen. The problem was not just that the meeting was poorly organized or that the Danish Prime Minister Lars Rasmussen was inept in this setting but that Sudan, at the time the head of the G77, set out consciously to undermine negotiations because Amnesty International had asked Denmark to arrest Sudan's President, al-Bashir, over human rights violations. Flannery thus came to realize that this conference would not result in a meaningful treaty on climate change. The political obstacles were simply insurmountable. Anthony Steinbock, talking about the phenomenology of hopelessness, has argued that the latter is 'the immediate and distinct experience of the impossibility of the event *as* impossible' (2007, 442). That is, hopelessness follows on the felt recognition that the desired outcome, narrowly specified, is impossible. *These* negotiations will not yield the hoped-for fruit.

By the end of the conference and for some time afterward Flannery describes himself as being 'exhausted' upon realizing that 'something fundamental had shifted' in our ability to tackle climate change through negotiations like this. More particularly, he came to believe that after fifteen years of trying to solve the problem, the U.N. was not the appropriate body to accomplish this. This insight goes well beyond the claim that particular countries, like Sudan or Canada, have in the past been mischief-makers in climate negotiations. It is the more radical claim that because of the structure of the U.N. and the pride of place it gives to nation states there is a failure of collective action at work here that is probably endemic to such enterprises. Now demoralization sets in. It involves the recognition that the forces arrayed against the hoped-for outcome are more firmly entrenched or intractable than one had at first believed them to be. Thus, whereas he had been inclined to think that the failure of Copenhagen was merely contingent, Flannery now sees it as the manifestation of a larger systemic problem, one that cannot be solved merely by changing the cast of characters involved.

Still, Flannery does not report thinking that we have been utterly defeated by the problem. The demoralization is still *relatively* specific. One way to understand this is to suppose that demoralization comes in degrees. At one extreme, one might believe that given our present political inertia we simply will not solve the problem in time to avert disaster. This would constitute total demoralization about this issue, but Flannery has clearly not come to this point. Further, the analysis reveals that demoralization is distinct from despair. Despair involves what Steinbock calls the 'loss of the ground of hope as such' (2007, 446). In despair the exhaustion that demoralized people often feel and that Flannery reports himself feeling becomes total. It extends to all our projects. Despair is more akin to nihilism than to demoralization.[8] But because his demoralization was not total and he did not succumb to despair, Flannery was able to renew his hope for a solution to this problem.

Imaginatively charting Flannery's psychic progression helps us distinguish hope from optimism, wishful thinking, hopelessness, demoralization, and despair or nihilism. Where does this leave us on the question of how we should hope for a solution to the climate crisis? In light of the numbers I cited in Chapter 3, achieving global decarbonization in the timeframe required to avoid catastrophe is, though clearly important, highly improbable. As I have already suggested this is therefore a case of hope against hope. But because of the strong claims of justice people of the future have on us we have a duty to hope for this outcome and to work diligently to achieve it. As I have argued, our recognition of the duty—the moral reason—can sustain the hope. Indeed without the sustaining moral reason, the hope will fade to a wish and is unlikely, as such, to be robustly motivating. In this case we will either sink into apathetic retreat or do the wrong things. Because of this our wishful thinking is also apt to degenerate into despair, as we notice that, while the disasters accumulate, the bad is gradually crowding out the good.

Like confidence or optimism, and unlike wishful thinking, rational hope is evidence-sensitive. To hope for an outcome is therefore to be open to the possibility that events might render that hope groundless. But if this is right, a hope cannot easily be separated from its corresponding fear. To hope rationally that p is, at least implicitly, also to fear that not-p (and sometimes, though not always, vice versa). So one way to understand our hopes better is through clarification of our fears, in this case our fears about the future after runaway climate change. In the next two sections, I'm going to set aside the concept of hope and concentrate in the first place on two possible futures that might await us: the first is dreadful (6.3), the second merely bad (6.4). After that, I'll return to hope (6.5 and 6.6).

6.3 The Moral Doomsday Hypothesis

Literary fiction can be a rich source of clues about possible futures. Consider in this vein Robert Wright's, *A Scientific Romance*. After discovering and repairing H.G. Wells' time machine, the novel's protagonist David Lambert travels to a future London inundated by climate change–induced sea rise. Equipped with little more than a collapsible kayak and a case of good rum, and paddling into what used to be Scotland, Lambert finds ubiquitous archaeological evidence of a world that had gradually degenerated into social chaos and violence, even cannibalism, as people scrambled to survive in the face of increasingly scarce resources and group fragmentation. To his horror, the tribe he ultimately locates—'an austere, superstitious, communitarian' people (Wright 1997, 226)—is racist, hierarchical, patriarchal, illiterate, in a state of constant readiness for war with other tribes, with a life expectancy of about forty years, and devoted to a strange and violent hybrid of Christian and folk-African ritual. Because his skin is lighter than theirs they decide he is the Messiah and attempt to crucify him in their annual passion play.

Lambert escapes this fate, finds his time machine, and returns to the age whence he came, thus repudiating a future in which, to understate the point, the bad outweighs the good. This is the sort of supplement we need to Lovelock's more ambiguous vision of the future. The key to Wright's novel, for our purposes, is Lambert's judgement that he is normatively *cut off* from these people and by extension from the entire future. Well before fleeing his ritual execution he expresses moral disgust with their values. Although he does not put the point this way, he comes to believe that the way benefits and burdens are dispensed in this society is fundamentally alien. It is not exactly unjust but *beyond* justice. Lambert's disgusted return to his temporal home has nothing to do with nostalgia—when he bolts the present he is a lovelorn, underemployed alcoholic—but rather points to a fundamental breach in human history. It indicates that there are worse outcomes for the species than extinction. In the course of his travels Lambert befriends a black panther whom he names Graham. Had there been no humans at all in the new world, Lambert might have stayed and lived out his days peacefully with his new companion.

I have been arguing that we are creatures of the historically specific drive to establish collectives that are consciously organized so as to minimize the rule of social or natural lotteries. This minimization is equivalent to attending to the vital interests of all genuine moral subjects. To have given over the dispensation of social benefits and burdens entirely to chance or official caprice effectively cancels *our* history. There is no reason to dismiss this as a piece of anthropological arrogance, as though only our values can matter. That kind of arrogance is objectionable precisely because in a world of diverse codes, many of which meet minimal standards of decency, it sets one above the others. Wright's novel, by contrast, portrays a world in which the only extant human group has organized itself around thoroughly indecent values (Lambert has no reason to believe there really are other tribes or, if there are, that they are any better). When he contemplates this world Lambert is prone to fits of despair about how we managed our collective way through the climate crisis.

The novel forces us to ask some troubling questions. Why should the bare persistence of humans into the deep future matter to us? Why should we care about it any more than we do the bare persistence of any other species? Martin Luther King Jr once said that the arc of history bends toward justice. This need not be interpreted as a claim about the way history inevitably goes, but a hopeful claim about how ours is going. But if we came to believe justifiably that justice would ultimately vanish from the human scene, though humans themselves would live on, it seems likely that all our efforts to establish just social relations now would also dissipate. This is because much of this work is motivated by utopian impulses. In this regard the work of and for justice is much like the search for a cure to cancer. Those working assiduously for the cure know that they are building on a research programme that may not bear fruit until well into the future. Similarly, those working on behalf of marginalized social groups might persist in their efforts even if they believe they will not succeed in this generation or even the next (Scheffler 2012). But in either case, if people came to believe that there was no future for their cause, what would be the point of continuing such work?

Let's call this outcome the moral doomsday hypothesis. In *Death and the Afterlife*, Samuel Scheffler asks how we would respond to two distinct literal doomsday hypotheses. The first involves an asteroid set to strike the planet in thirty days, thus destroying all of life. The second is from P.D. James' *Children of Men*, a story in which the species has become sterile. Scheffler is interested in what would happen to our sense of our own projects in such a world, especially the second. He argues that we are more dependent on the 'collective afterlife' (the continued existence of future people) than we recognize. James' protagonist, Theo, makes the key claim: 'Without the hope of posterity . . . for our race if not for ourselves, without the assurance that we being dead yet live, all pleasures of the mind and senses sometimes seem to me no more than pathetic and crumbling defences shored up against our ruins' (Quoted in Scheffler 2012, 41). Scheffler thinks that Theo is right, that 'the prospect of our imminent extinction would occasion widespread apathy, anomie, and despair' (2012, 39). The afterlife matters to us in its own right but mostly because it is a condition for anything else mattering to us, a fact that highlights our tendency to 'personalize our relation to the future'.[9] There is thus an ineliminable generational conservatism at the core of many of our values, which is why Scheffler says that 'we need humanity to have a future for the very idea that things *matter* to retain a secure place in our conceptual repertoire' (2012, 60).

In Chapter 4, I talked about justice partly from the standpoint of the main victims of climate injustice—future generations. The reason for this is that people of the future clearly depend on us and justice is paradigmatically about securing appropriate benefits and avoiding inappropriate burdens for moral subjects. Justice in the soul also has this focus: those possessing the virtue of justice are uniquely attuned to the way their actions affect the vital interests of moral subjects. But what Scheffler's arguments show us is that we are as dependent on people of the future as they are on us. As we have seen, this is why believing that the species is about to go out of existence can hollow out the commitment to some of our deepest values. But this is surely just as potent a threat in the case of the moral doomsday as it is for Scheffler's literal doomsday. The anomie and despair occasioned by our belief in the literal doomsday hypothesis would, I suggest, be just as powerful were we convinced that a moral doomsday looms.

If this is right, it shows that we are more pervasively concerned with justice than we might have realized. In our political culture, naked appeals to group or sectional interests are difficult to sustain. Despite significant backsliding, inclusivity is increasingly the norm. There is of course a danger that much of this concern is superficial or ideological. Indeed, if ideology consists in presenting the partial or sectional as the universal, our concern for inclusivity opens up significant space for ideological distortion. But the ability to invoke concerns of justice, even insincerely, depends both on our having the concept and seeing that appearances *sometimes* match reality. This is one way to define the work of justice: to bring appearance and reality ever closer in this domain. But the key point is that this work is also essentially diachronic (and also narrative, as I will argue in Chapter 7): it would stop dead in its tracks were we to come to believe in the inevitability of Lambert's world. In this case the work of justice might not be replaced with utter lethargy but, perhaps even more disturbingly, with acquiescence to group fragmentation. If the latter is ineluctably our lot, we may as well start picking sides now and arm ourselves for the coming troubles.

So it seems difficult to deny that we have a *stake* in how the future unfolds. In light of the climate crisis, it has been tempting for some to say that those of us pushing hard for intergenerational justice are trying to constrain the future in undesirable ways or to an undesirable degree. Those defending free market approaches to the problem often make this kind of claim. Human ingenuity is, so goes the claim, boundless, so let's just leave the future to people of the future. We even hear that there is evidence for this optimism in our deep evolutionary history. After all, humans moved out of Africa and populated the rest of the planet in the midst of ice ages and other environmental dangers.

If they could do it, surely our descendants will be no less ingenious at contriving ways to survive in the teeth of their own environmental challenges.[10]

There may be an assumption in claims like this that so long as the species survives, so too will everything else that really matters to us. The thought may be that the sense of justice, for instance, is innate in the species and that even if it is extinguished for a time it will inevitably be rekindled. But there is no reason to believe this. The sense of justice we have and are in the midst of developing is a fragile and entirely contingent thing. People of the future may evolve cultural worlds that are from our current perspective unimaginably strange, and there is no reason to bemoan this. But if they lose the sense of justice a key connecting thread to us will be broken and *they* will be unrecognizable as a result. They will, I submit, no longer be our descendants whatever our shared biology suggests to the contrary. Though it may sound unduly proprietary, to this limited extent the future is either recognizably *ours* or it is not a human future worth caring about.[11] We therefore have an interest in the collective afterlife but, with due respect to Scheffler (and Theo), this afterlife must involve more than the mere existence of future people.

6.4 Triage and the Survival Lottery

This brings us to our second possible future. Much in the present analysis hangs on what is meant by the 'sense of justice'. Again, justice as I understand it has to do with diminishing the social or institutional power of chance and attending to the vital interests of all moral patients in the dispensation of social benefits and burdens. This ideal can be preserved both dispositionally in agents and partially in institutions, even if circumstances do not allow for its full expression. There are two possibilities to consider here: triage and the survival lottery. Triage is the practice of distributing vital benefits and burdens on grounds of efficiency alone. Because by hypothesis this is a world of extreme scarcity, each person will be assessed, as Catriona McKinnon puts it, on whether or not she is an 'efficient convertor of resources into life'. But, clearly, one's ability to do this is a matter of chance:

> Possession of this property in most circumstances is a matter of brute luck: a person's genetic inheritance, her age, her upbringing, her place of birth and living, her diet, her level of education, her exposure to natural disasters and toxins in her environment, etc. all bear on the extent to which she is capable of converting resources into life... [F]or principles of triage to be defended as principles of allocative justice requires a defence of the moral significance of possession of the property 'being an efficient convertor of resources into life' fit to justify the outcome of death for those who lack it, or have less of it than others. I submit that no such account is available. (McKinnon 2012, 120)

The situation becomes even more taxing where there are competing claims for resources among those equally endowed with the key property. Here, according to McKinnon, the only thing to do is institute a survival lottery. Tim Mulgan posits an identical outcome in a climate-ravaged future world. Here is how the imaginary inhabitants of that world (the 'broken world') say they would describe it to a fictional present person (a denizen of the 'affluent world'):

> The most striking feature of our societies, for an affluent visitor, would be that survival bottlenecks are an ongoing fact of life. We often find ourselves in a place where we cannot all survive. The central questions of our political philosophy are: how do we preserve society through these bottlenecks; and what do justice and ethics require in such extreme circumstances? (Mulgan 2011, 10)

From the standpoint of the present analysis, Mulgan's construction of the problem is especially pertinent because he puts members of the two generations into imaginative contact and asks how they might understand one another. Members of the affluent age could, he suggests, see a connection with these people because there are two places where survival lotteries took place historically:

> Before they were integrated into affluent society, the Eskimo practiced both infanticide and euthanasia as ways to ration food in harsh times, as did many other pre-affluent peoples. And every affluent society faced the problem of distributing comparatively scarce life-saving treatment. The pace of technological 'improvement' was never going to be sufficient to keep everyone alive forever. (Mulgan 2011, 11)

These two examples are importantly different. In many societies, one is far more likely to receive life-saving medical treatments, like affordable drugs, if one is already a member of some favoured social group. Such treatment is a social good that is often distributed on the basis of economic or social privilege. The way the 'Eskimo' dealt with the old and infants in times of extreme scarcity is not like this. Here, the survival interests of the collective were paramount. So we should distinguish between exclusion from social goods based on privilege or official caprice and exclusion based on commitment to collective goods, the extreme instance of which is group survival. The key point is that the latter, but not the former, *can* express the value of impartiality.

But with respect to this value triage and the survival lottery are importantly distinct approaches to allocative justice. A survival lottery can still count as an exercise of justice because, barring corruption in the process, it is constitutively impartial. Each person, for instance, would get a single ticket each of which has an equal chance of being drawn in a transparent selection procedure (McKinnon 2012, 126). However, McKinnon is surely correct to argue that possessing the property 'efficient convertor of resources into life' is a matter of pure chance. Even if triage is supplemented by a lottery, as when there are equally efficient competitors for scarce goods, the procedure still excludes those who happen to fall outside the circle of efficient agents. This, alas, is unavoidable. I see no way to deny the claim that efficiency at converting resources into life must be a baseline for decisions about allocation of resources in times of severe scarcity.[12] The best we can do is to insist that where competing claims exist among equally efficient agents an impartial lottery should be run. Because the lottery (which can be just) is run on the back of triage (which is not) the two-tiered process is a mere shadow of justice. But there's some substance in shadows, especially by comparison to some of the alternatives (like Lambert's world).

At this point, we need to emphasize a point which is insufficiently stressed in the scant literature on this topic. Any actual survival lottery might have a blind spot with respect to the demands of intergenerational justice. Although the point of the lottery is to allow the group to survive for a little longer, there is no guarantee that any particular instantiation of it will ensure that the interests of far future generations are taken account of. After all, those deemed most efficient in their generation, who also, let's suppose, come out on top in a fair lottery, might decide to hang the next generation altogether and use up the remains of this or that scarce resource. This would be deplorable but would not obviously contradict the principles of justice underlying that instantiation of the lottery because this had to do *entirely* with the distribution of benefits and burdens among existing people (who lives, who dies). To avoid this, no running of a survival lottery can be confined to members of any present generation. When facing a population bottleneck, members of a present generation should include future people in its deliberations about who should and should not get life-saving resources, effectively issuing survival tickets to those people. How far into the future this consideration should extend is difficult to determine a priori. There are First Nations communities in North America for whom 'each new generation is responsible to ensure the survival of the seventh generation', and this strikes me as a model of intergenerational responsibility eminently worthy of emulation, even (or especially) in times of crisis (Clarkson et al. 1992).

This approach is committed to treating as equals anyone capable of efficiently converting resources into life. But how could we deny that any particular member of the next (or the seventh) generation will have this capacity to a degree equal to or exceeding that of people of the present? We can fix it so that people of the future lack the capacity for efficiency, by using up or degrading the material means necessary for it. However, we are supposing that we cannot justify begging the question in favour of our own survival in this fashion. Since in advance of enacting a particular policy we cannot know which future people will be efficient convertors and which will not, the commitment to impartiality must be extended to some subset of future people *as a whole*. We, members of *any* present generation, must assume that people of the future have a claim to scarce and essential resources equal to our own, or more precisely those of a subset of us, and 'issue them tickets' in the lottery. The number of future people we are talking about need not be so large that their claims swamp those of the present generation. A generation in the midst of a population bottleneck need think only about what is minimally required to keep the group going down the generations (as opposed to trying to expand the group's numbers, for example).

Though we may face literal survival lotteries in the future these considerations also point towards the more general importance of sustainable resource management in the present. Again, if we seek guidance and inspiration on this issue perhaps we should look more seriously at the traditional ecological knowledge of the world's aboriginal communities. After all, they have been honing the practice of taking the far future seriously for many thousands of years, often in the teeth of severe population bottlenecks induced by adverse environmental conditions.[13] Obviously, the pressure on presently available scarce resources becomes more acute the further our future concern is cast. Where population bottlenecks are especially tight and we are also concerned to persist into the far future, many members of the present will not survive. At the moment, we are entirely too complacent about these possibilities. So wedded are we to the idea that our collective affluence can grow ceaselessly that the prospect of our one day having to make hard choices like this is beyond the pale. That is a dangerous illusion, which is why we must begin thinking much more seriously about the hard discipline of sustainable living. If we do not do this we will have little chance of managing future resource conflicts justly. If we look for comparison to the experiences of the seventeenth-century (examined in Chapter 1) we see that the failed adaptation to climate change had much to do with poor management of population bottlenecks. The sixteenth century had been unusually warm, which allowed for an expansion of populations across the globe, populations that crashed in drastic ways when the weather changed and governments scrambled to adapt. Again, Japan seems to have been the lone exception, largely because of its 'demographic cushion' (Parker 2013, 488–97).[14]

Is it morally outrageous to suggest that we might govern ourselves this way? Again, on the sort of scheme I am talking about, it will be possible to rescue even fewer moral patients from any current generation than would be the case were the far future not brought into that generation's deliberations about how to get through a bottleneck. But this is a difference of degree, not of kind. Any system of triage and survival lottery involves sacrificing the interests of some moral patients in the interests of allowing the group to persist for a time. If the principles underlying these practices are morally acceptable at all, then it is difficult to see how we could refuse to extend them in the way I am suggesting without inviting the charge of moral arbitrariness. And the extension is justified, as I have said, because it forestalls the possibility that any set of generational 'winners' might think itself entitled to go out with a bang. In the end, extension of the lottery to the far future—for example, leaving future people some coal and oil in a world that has, perversely, not moved beyond its dependence on fossil fuels—is simply a demand of fairness. As Shue puts a related point, it makes little sense to say that we will do what is fair, 'except for the intergenerational part' (2014, Chapter 16).

The two futures I have sketched in this and the previous sections are not mutually incompatible in the sense that if one emerges, the other cannot or will not. In particular, it is quite easy to see how the less fearsome future—triage plus survival lottery—is terribly precarious and might therefore give way to the more barbarous outcome in a single society. Also, the world might be dominated by societies employing triage and the survival lottery, with pockets of barbarism here and there. Or vice versa. My attention in this section has been focused on the way in which people of the future might seek to live justly, even as they fight through the damaging effects of climate change. Hope will clearly help them and they will no doubt seek to sharpen this hope by contemplation of the alternative. That might make them especially vigilant guardians of their survival lottery's integrity. What about us? What should our hope look like in the midst of this crisis?

6.5 Hope in the Anthropocene

Jonathan Lear has given us a picture of the Crow people at a time of profound cultural upheaval for them, the catalyst for which was the near-elimination of the buffalo. As the tribe's leader Plenty Coups put it, after the buffalo were gone 'nothing happened' in the cultural life of his people. And yet Lear argues that Plenty Coups brought the Crow through this period with an attitude of 'radical hope'. I will have more to say below about what this attitude consists in, and will argue further that it is the sort of hope we should adopt for the climate crisis, but for now I want to highlight the more general point Lear is addressing, namely the possibility that any of us could experience a similar collapse:

> If this is a human possibility, philosophy—in its ethical dimension—wants to know: How ought we to live with it? So: it is one thing to give an account of the circumstances in which a way of life actually collapses; it is another to give an account of *what it would be for it to collapse*. And it is yet another to ask: How ought we to live with this possibility of collapse? (2006, 9)

As we have seen, many people writing about climate change express the hope that complete social collapse can be averted. Lear thinks that for Plenty Coups radical hope was the alternative to resignation in the face of imminent collapse: 'What makes this hope radical is that it is directed toward a future goodness that transcends the current ability to understand what it is. Radical hope anticipates a good for which those who have the hope as yet lack the appropriate concepts with which to understand it' (2006, 100). On Lear's understanding of his achievement, Plenty Coups hoped to preserve Crow *agency* in a world in which they could no longer do most of the things they had traditionally done, many of which were of central cultural significance. But he did not know exactly what that agency would look like. Nevertheless, seeing that the onslaught of white settlement was likely unstoppable, he made peace with the American government in exchange for guarantees of some protection of traditional Crow lands under the reservation system. To many in his own tribe as well as other tribes—the Sioux for example—this looked like a sell-out. To come to our case, why is the hope that we will rapidly decarbonize the global economy best thought of as radical, as I have just hinted is the case? The key here is to distinguish between instances of hope whose objects differ in degree of determination. How indeterminate must the object of one's hope be in order for the hope to count as radical?

Clearly, the goal 'rapid decarbonization of the global economy' is relatively determinate. We have fairly precise numbers about temperature anomaly ceilings, carbon usage by Gigatonnes (Gt), timelines, etc. This is so even though there are multiple paths that might realize the goal. For example, the goal of decarbonization doesn't by itself tell us whether to implement a carbon tax, a cap-and-trade system of carbon permits, some hybrid of the two, or some fourth option. But as we will see in Section 6.6, rapid decarbonization demands such sweeping transformations of our culture that we find it difficult to imagine what the latter might look like on the other side of the transformations. So beyond the specifics of our energy regime, the society we are hoping for is indeed indeterminate. Even with respect to energy, things are not very determinate: we know exactly what we don't want (continued reliance on carbon-based energy sources), but much less about exactly what we do want (more nuclear?).

Even so, it might be suggested that the object of the hope I have identified is too determinate to qualify as radical. Plenty Coups was, after all, talking about the seemingly radically indeterminate hope that something 'good' would eventually transpire for the Crow. The key point about a radically indeterminate hope, so the argument might go, is that it cannot be disappointed, whereas it is pretty easy to see how the hope I have identified for our crisis might be. Martin interprets Plenty Coups' hope this way:

> His hope was such that it would always be rational for him to deny that any particular worldly outcome was its disappointment, or that any experience was

evidence that the outcome for which he hoped was not good. Plenty Coups was thus justified in adopting the meta-confidence of faith, the attitude that no experience could provide reason for him to stop seeing the chances of his tribe surviving and flourishing as sufficient to license hopeful thoughts, feelings and plans. (2014, 106)

Martin fails to explain how the justification works here, and this failure exposes a key problem in her account of hope. Much of the literature on hope, as we have seen, is very careful to distinguish it from wishful thinking and self-deception. Martin herself is at pains to show that the hopeful agent's subjective probability assignments must track the truth. She gets into trouble because she is eager to legitimize faith as a species of hope, seeing a connection between the theological hope of Gabriel Marcel and the quasi-secular hope of Plenty Coups. There is no need to deny that this sort of hope is often 'sustaining', but given her insistence that subjective probability assignments must be truth-tracking, Martin has not shown that it is 'a *rational* form of confidence in the possibility of a good outcome to a trial' (2014, 117; my emphasis).

It is in any case a forced interpretation of Plenty Coups, who must have had *some* concrete notion about what the good for the Crow consisted in. After all, he argued vigorously against both allotting land to the Crow on an individual basis and forfeiting mineral rights to retained land (Lear 2006, 137–8). How could he have done this without relatively specific beliefs about what is materially necessary for his people to live good lives? So he must have had some notion of what it meant for the Crow to flourish *as* the Crow. If I am right about this there are, *pace* Martin, some outcomes that would defeat Plenty Coups' hope. Presumably, for example, allotment of land on an individual basis would have corroded the life of the collective to such an extent that it would be impossible for Crow culture, and therefore Crow agency, to thrive in such circumstances. To sum up all of the concrete practices—as well as their material preconditions—deemed by Plenty Coups to be necessary in this way, I suggest, just is to define what is meant by the good for the Crow. I see no reason to believe that Plenty Coups thought of the good in total abstraction from such practices and their material preconditions. I conclude that radical hope is determinate enough to allow for its defeat. Hope's rationality can be secured only by *distinguishing* it from faith in this way.

Still, Martin's interpretation has merit because it marks a clear distinction between radical and non-radical hope: the former is non-defeasible, the latter defeasible. Since I have argued that all hope is defeasible, it might be suggested that by calling some instances of hope radical and others not I am merely applying a label. This charge can be answered by noting the connection between radical hope and revolution. Radical hope is intrinsically minimalist: it recognizes the inevitability of significant loss but focuses on that which can be preserved through the tumult. It is therefore also necessarily revolutionary.

Further, although all radical hope is hope against hope—the minimalism implies that many of the most important things are deeply threatened—the converse does not hold. The hope of the person with late-stage cancer may or may not be radical. If she dreams of a future smoothly continuous with her life before the diagnosis it is not, but it *is* if she thinks that on the other side of her struggle virtually everything must change for her. The conceptual connection to revolutionary tumult thus allows us to distinguish between radical and non-radical hope. What we are trying to preserve across the generations is the sense of justice. It is not only that achieving this is compatible with everything else being up for grabs, but that revolutionary change is in fact necessary given the penetration of the fossil-fuel regime into every aspect of our lives. What prevents us from seeing this clearly and acting on it?

6.6 Presentism and the Grip of the Past

Implicit or explicit calls for revolution are ubiquitous in current climate change literature. We are regularly informed that we need to overturn or reform radically our attitudes, economics, energy policy, political institutions, social relations, and more. Consider this representative appeal:

> The reality is that the major environmental problems we face today—of which climate change is only one—cannot be solved by means of technological or market-based solutions while keeping existing social relations intact. Rather what is needed most is a transformation in social relations: in community, culture, and economy, in how we relate to the planet. What is needed, in other words, is an ecological revolution. (Magdoff and Bellamy 2011, 112)

In a similar vein, Philip Cafaro has argued that 'meeting the global climate change challenge almost certainly depends on ending human population growth and either ending economic growth or radically transforming it' (2011, 212). Stephen Gardiner, for his part, suggests that we probably need to 'create new institutions' in order to deal effectively with climate change (2011, 55). Finally, Henry Shue argues that we should see our current situation as part of a three-revolution historical arc. The first revolution—the Agricultural—allowed for the creation of large-scale human civilizations centred in cities. The second—the Industrial—enabled the vast extension of that civilizing project through the harnessing of fossil fuels. However,

> The second revolution, we now realize, threatens to undercut the first. The third great revolution must, therefore, be the creation of both an escape route from fossil fuel energy and a path to the most rapid possible transition to alternative sources of energy in order to preserve the ecological preconditions for sustainable development. How should we make this great transition from the carbon-based fossil

fuels with which we are now cutting our own throats to safer sources of energy? (Shue 2013, 385)

In this section I want to try and uncover the philosophical roots of our resistance to this idea, not in the abstract but as applied to the 'great transition' Shue is talking about. Revolutions are intrinsically future-directed phenomena. When we make a call for a revolution—any kind of revolution—we are asking for a 'fundamental change or reversal of conditions', as the *OED* has it. In the name of the future, revolutions are conscious attempts to abrogate the privileges of a present too tightly bound by the traditions and practices of the past. In the analysis of Scheffler we have seen that some conservatism about the future is inevitable. For us to care about our collective projects in the right way or to the right degree we assume some level of future-projected normative continuity. We might, however, distinguish conservatism thus construed from 'presentism', the belief that the suite of normative commitments and institutional arrangements that make up the present, themselves the relatively unbroken inheritance of the past, fully define future possibilities.

Here are two important examples of presentist thinking. First, note Shell's ex-CEO John Hofmeister's description of the 'bold' forward thinking of top executives in his industry. This is from his recent book, *Why We Hate the Oil Companies*:

> [O]il executives . . . focus on what they believe is the best thing they can be doing to benefit the consumer: finding more oil and gas supplies for the future. The oil that will be in gas pumps and cars and planes and ships for the next eight to ten years has already been found. Oil executives are looking beyond that time frame to the resources that are still to be discovered or tapped . . . The future is now for every oil company. The clock never stops ticking on the need to identify and access future supplies, or reserves. It is a compelling, ongoing obsession for top leaders. (2010, 132–3)

Hofmeister is deeply impressed by this future planning. If there's a problem with the oil industry, he thinks it's merely that such a focus by executives tends to put them out of touch with the dirty details of the retail market. People 'hate the oil companies', we are informed, mainly because the restrooms in filling stations are always such a damn mess. But oil and gas are, though 'unlovable', also 'unavoidable', so we had better get used to their dominance in the energy mix for decades to come. I have no doubt that Hofmeister has accurately depicted how oil executives think and operate. What is shocking in this account is how deeply entrenched the presentism is.

Although Hofmeister's book is in large part ostensibly about the future, there is no serious attempt to grapple with the scientifically established connection between the enhanced extraction and use of fossil fuels and worsening climate change. Not a word about the welfare of future people outside the specific context of the demand for gas and oil, which by his own account his industry is striving to keep vital for as long as possible. While we may be stuck with fossil fuels for some time, two points must be noted. First, this situation is largely the product of a century or more of externalization of environmental and social costs on the part of the oil industry as well as the market-distorting subsidies the latter have received from governments. Among other things, these factors have resulted in the severely retarded development of renewables, a supply that would have been more robust by now if not for the dominance of fossil-fuel interests in our economy and political system. Our dependence on these products, in other words, has been artificially constructed and maintained.

But second, it is *not* necessary for us to extract and burn more of this stuff than a responsible carbon budget allows. As Hofmeister puts it, 'the most useful energy source we have for now and the near future is hydrocarbons. The world's use of oil and gas and coal is going to intensify over the next 20 to 50 years or longer' (2010, 72). Hofmeister thinks the 'waste' this use produces—carbon emissions—should be 'managed' the way we manage other wastes, through trading permits and sequestration, for example. He does not even contemplate imposing a serious constraint on supply through, for example, a carbon tax. He does mention the need for a 'cap' on industry emissions. But he gives no details about where the cap should be set, neglecting to state that it should correspond strictly with what science tells us we need to do. And since he welcomes the 'intensification' of fossil-fuel use over the next half century *or more*, we can only assume that the cap he has in mind must be located somewhere in the stratosphere.[15] What's behind all of this, of course, is a refusal to advocate measures that would to any degree threaten the affluence and high-consumption lifestyles of the developed world (especially in the U.S.).

Hofmeister styles himself a straight-talking insider, intrepidly challenging the easy assumptions and complacencies of his former colleagues. This appears to involve nothing more than the bland suggestion that *someday* fossil fuels will be replaced and that the oil companies need to be clear-eyed about this. Within the oil industry this is as critical as thinking gets. The notions that we might (a) act *aggressively* to constrain the supply of fossil fuels (i.e., set the ceiling where the carbon budget, rather than the oil executives, dictates it should be) and (b) divert the lion's share of what remains of this energy resource to those who really need it (those in the developing world) are not even on the horizon of possibilities, and therefore the crisis we are in cannot so much as be perceived. Nor, therefore, can alternative futures.

But it's the second example of presentism I want to focus on, both because of its generality and since it returns us to Gardiner's seminal way of framing these issues. Think of how national governments negotiate emission reductions targets, the 'trajectory' issue. As Gardiner points out these negotiations are plagued by invocations of 'prior entitlement':

It is standard to frame climate policy in terms of national targets based on each country's emissions in some base year . . . But this is to embed historical patterns of emission. It assumes, in effect, that countries with historically higher emissions begin with some kind of entitlement to that higher level that those with historically lower emissions do not have. (2011, 317–18)

What I want to examine here is the dangerous way entitlements 'embed historical patterns of emission'. Rhetorically if not conceptually, an entitlement is a necessity. In using this language, developed countries are clearly protecting their luxury emissions. Since these emissions will be protected for as long as countries can manage to do so, the really worrisome possibility is that we will set in motion what Gardiner calls an 'intergenerational arms race' (IGAR) (2011, 185–207). To the extent that we believe we are bound to experience climate-induced disasters, we will likely relax efforts at mitigation and focus exclusively on adaptation. And even though some adaptations will benefit future generations, many will simply protect luxury emissions: 'the current generation will have an incentive to prioritize projects and strategies that are more beneficial to it (e.g., temporary "quick-fixes") over those that seem best from an intergenerational point of view' (Gardiner 2011, 200).

Embedding historical patterns of consumption in this manner might lead inexorably to the radicalization of our generation-relative stance. The pure generational problem (PIP), recall, involves passing climate change costs down the generations. The new and more radical problem has to do with the way these costs mount for each successive generation. Each will burn as much fossil fuel as it deems necessary to cope with its own problems, invoking the entitlement to self-defence to justify its actions and exacerbating conditions for every succeeding generation. The problem is iterated; the harms visited on each successive generation increasing cumulatively. By hypothesis, in the age of the IGAR, we do not think of ourselves as constrained to any degree by consideration of the vital interests of future generations. This represents a dramatic narrowing of our temporal horizons, which, in turn, implies belief in a constriction of our agency that is likely to induce fatalism. Concretely, this means we will come increasingly to believe that we cannot fundamentally alter the energy regime with which we are saddled. In this area at least the past has us fully in its grip. This is presentism in spades.

The only way out of this is to challenge the rhetoric of entitlement as it is currently deployed in climate change negotiations. Again, the key feature of this concept is its connection to a claim of necessity: we *need* to maintain or increase our historical pattern of emissions.[16] However, as Judith Shklar has argued, appeals to social, political, or economic necessity are 'the staple item of ideological discourse everywhere' (1990, 74). They are generally meant precisely to obscure the degree to which societies and governments *do* have significant control over the choices they make. As I have argued in Chapter 2, we are free to do what morality requires in this case, though ideology functions to conceal our freedom. I've said enough already about how this applies to us now, but notice that it also applies to a future where the IGAR is operative. In the throes of the IGAR the ramped-up use of the fossil-fuel stock is justified by appeal to self-defence, as we have just seen. Now, it is notoriously difficult (though not impossible) to justify harming innocent third parties in the course of defending oneself. But in the case of the IGAR the aggressor is the previous generation while the (unintended) target is the next one. This places in doubt the moral justifiability of the self-defence option in the IGAR.[17]

The shrillness of appeals to entitlement and necessity will doubtless intensify in that world, but the thought process just sketched will nevertheless always be available to anyone willing to articulate it. And to articulate and act on it—by refraining from bringing harm to next-generation innocents—is an exercise of freedom. Members of any generation faced with the sorts of choices that define the IGAR can decide to sacrifice some members of their own generation in the interests of bettering the future, or at least not making it worse. They can, that is, issue tickets in the survival lottery to people of the far future and thus refuse, to some degree, to 'take up arms' in their own defence. Where dependence on fossil fuels persists this would involve constraining consumption of them for present-focused adaptation. On the assumption that there is no other way to cope with present disasters, this means that some people who could have been saved will not be. It is important to point out that such a momentous decision connects us to these people in a deep way. We can see them as carrying on a project—the Enlightenment commitment to impartial justice—that is also our project.

Presentism may be ubiquitous in the history of societies. But when murmurs of revolution begin rising to the surface of a social order, the specific form taken by presentism there—expressed both in its social relations and in the souls of its people—can finally be picked out for what it is. These are clarifying historical moments, where the ideological fog seems finally to be lifting. This is where we are with climate change. Environmentalists sometimes talk about James Hansen's 1988 testimony to Congress about climate change's dangers as having happened so awfully long ago, and they bemoan the political inertia that has overwhelmed us in the meantime. It *is* truly shameful that we have done virtually nothing about the problem since then. But in the larger historical perspective it is just *dawning* on us that the many forces sustaining our current energy regime are enemies of the future of our species. This dawning, we might say, is revealing the extent to which our agency is unduly constrained under the alienating conditions peculiar to the age of fossil fuels and climate change. As I argued in Chapter 2, many of us increasingly feel the need to take the future back from these forces. Concretely, this means opposing the logic of entitlement before it festers into an intergenerational arms race.

Finally, let me emphasize that it is not enough to hope simply that we don't descend to barbarism. Divorced from the immediate task of decarbonization, this is simply not specific enough and is as such likely to degenerate into the vague wish for a miracle.[18] Again, our complex hope should be that we decarbonize the global economy rapidly *so that* future people are given the best possible chance to flourish and maintain or enhance the institutions of justice. We can connect the two parts of the hope because there are rough correlations here. The hotter we let it get, the more scarcity there will be for future people, and, other things equal, the more difficult it will be for them to sustain just institutions in such circumstances. All of this is consistent with admitting that we do not know at what precise temperature anomaly the bare struggle to survive swamps the attempt to organize societies on the basis of considerations of justice. Although it is possible that with, say, a 3°C anomaly things would not be so bad as to require extreme measures, we can't say for sure. Maybe at that anomaly we simply panic, perhaps worrying that we are on the brink of tripping some of the climate system's scarier positive feedbacks—like deep-sea methane clathrates—which would send the climate into runaway mode and human systems into chaos. Or maybe because of the resilience of our commitment to justice, or the emergence of other serendipitous contingencies, we don't get barbarism short of a 6°C or even 8°C anomaly.

The latter outcome is, we should emphasize, a mere possibility. We know that anything above a 2°C anomaly flirts with the possibility of moral disaster. Nor should this be seen to invite resignation insofar as we judge the 2°C target unrealizable (which it probably is). Given the rough correlations just cited, recognizing our ignorance of salient thresholds and contingencies is compatible with assuming that the more aggressively we decarbonize the better. There will always be opportunities for dissembling and evasion as we think through these things, but that fact cannot obscure the clear distinction between a sincere effort to decarbonize and its mere pretence. Decarbonization is not an end in itself. It is simply a (necessary) means of carrying the larger Anthropocene Project forward. Our task is to focus on—to hope for—the achievement of *this* condition, not all the other conditions that together are sufficient for bringing about a world in which people can flourish.

6.7 Conclusion

Plenty Coups' ability to imagine a new future for his people was revealed to him in a series of dreams beginning when he was a child. In these dreams he encountered unnamed but 'powerful and malevolent forces that cannot be contained', forces that were out to 'destroy a way of life' (Lear 2006, 129). The dreams were tapping into a generalized cultural anxiety felt by the Crow in the face of the increase in white settlers and the disappearance of the buffalo. They are not mere imaginings but calls to wake up to reality, something most members of the tribe do not initially want to do. Plenty Coups was able to go forward hopefully only because the tribe's elders took his childhood dreams as *authoritative*. Similarly, we will overcome our political inertia only if we begin to treat the near-unanimous findings of climatologists with the degree of authority they deserve. A crucial element of this strategy will be broad dissemination by our political leaders of the idea that past political and economic practices are, at best, imperfect guides to the policies we now require. This will require a new kind of leader as well as a new level of cultural seriousness about the socially relevant findings of science.

The mounting anxiety in our culture about this issue is a warning to us. Alexis de Tocqueville makes a claim about the French revolution that should give us pause in the present crisis:

> But of all the strange phenomena of these times, the strangest to us, who have seen so many revolutions, is the absence of any thought of revolution from the mind of our ancestors. No such thing was discussed, because no such thing had been conceived. In free communities, constant vibrations keep men's minds alive to the possibility of a general earthquake, and hold governments in check; but in the old French society that was soon to topple over, there was not the least symptom of unsteadiness. (2010, 163)

The rest of his analysis makes clear, however, that there was in fact a deep sense of grievance among the peasantry and the emerging bourgeoisie as well as the *philosophes* who gave literary expression to the general malaise. But the elite did not perceive 'the least symptom of unsteadiness'. The resulting upheaval was, de Tocqueville suggests, more terrible than it might have been if those in a position to change things had been better attuned to the anxiety in the rest of French society. Upheaval was probably inevitable, but it might at least have been channelled into less violent paths.

Something similar is happening now and it is crucial to begin the task of shaping its development justly so that we are not overwhelmed by it down the road. We need to reclaim our agency and this is why we must summon hope's constitutive forwardness, its way of incorporating a positive vision of the future into our present activities. Because it represents the reclamation of agency, hope shatters ideologically constructed appeals to necessity and the fatalism and despair they entrain. Instantiating the radical hope of rapid decarbonization requires working for meaningful political change, acting in newly courageous ways, and looking hard for alternative models of sustainable living. Because of its peculiar temporal outwardness, and working in concert with the other two virtues—justice and truthfulness—hope allows us to identify the enemies of future humanity, the defenders of the fossil-fuel industry. It is our job as hopeful citizens of the present to confront these people head on by challenging their unwarranted presentism. If we can summon the imagination, moral seriousness, and humility needed to do this we have a chance of succeeding. Hopefully.[19]

7

Conclusion

Will They Forgive Us?

7.1 Introduction

In her recent book on forgiveness, Margaret Holmgren writes,

> Human history is full of turning points. Although we may prefer not to think about it, the one we face now is arguably the most momentous of all that have occurred up to this point in time. If we do not learn to work together as a global community within the next decade or two, we may well cause irreparable damage to the planet that results in seriously diminished life prospects for all who inhabit it in the future. (2014, xi)

Since we are in the process of causing this large-scale damage, the question of whether we will be or should be forgiven for our wrongdoing is bound to arise. After this promising comment, however, Holmgren more or less drops the issue of intergenerational forgiveness in the wake of climate change. No other philosopher has seen fit to explore the issue. And yet, it looks to be a prime candidate for such an exploration because there is widespread agreement among philosophers that climate change does present us with problems of justice and wrongdoing, and that these issues have both international and intergenerational scope. That's all we need to get the relevant normative analysis of forgiveness off the ground, so why the hesitation? In this concluding chapter I will seek to answer this question, show that the hesitation is misguided, and provide an analysis of intergenerational forgiveness applied to the problem of climate change. The point of all of this is to show that the possibility of intergenerational forgiveness is a key component of building a meaningful sense of ourselves as participants in a multi-generational moral and political project. What I have to say here therefore ties the problem of forgiveness back to the larger themes of the Anthropocene Project. I begin with a look at how the concept of moral indignation might help us gain a fresh perspective on Parfit's non-identity problem.

7.2 The Non-Identity Problem

Why have we been hesitant to look at climate change from the standpoint of intergenerational forgiveness? Ultimately I will suggest that to do so is to admit that the present is wronging the future and that is a thought we are not yet ready fully to entertain. I don't mean we have good arguments about this, just that we can't face the wrongness squarely. But there are other possibilities—including our having good arguments—so we had better look at these first. It might be that there is scepticism about the extent to which it makes sense to forgive the dead in general. But this is surely unwarranted. Interpersonal forgiveness is, paradigmatically, the forswearing by a victim of negative emotions—resentment, anger, indignation—directed at the person who has wronged her. Because the focus here is on what the victim decides to do, there is no reason in principle why the process cannot take place with respect to dead wrongdoers. In the interest of moving on with her life, an abused daughter might decide to forgive her abusive parent, even though the latter is now dead and never repented of his crimes while alive. Or maybe the perceived problem has to do with how a collective can forgive another collective. It is after all difficult to know how to attribute emotion to a collective agent.[1] However, although I will be—and have been—speaking of what future generations might think and feel about our generation, I'm really interested in the way individual future agents might shape their reactive attitudes towards us. There does not seem to be anything philosophically outlandish about such an approach, so I don't see how this could be the source of the hesitation I have identified.

Maybe we need to take a step back and ask whether or not future generations can rightly blame us for what we are doing or failing to do. It may be thought that the negative effects of our actions and omissions are so temporally far-flung that the people 'victimized' by them would not have existed but for these actions and omissions. The original events or decisions become so causally refracted through time that they result in different people being born than would otherwise have been born. And if this is correct, then, so long as their lives are, on balance, worth living, they cannot rightly reproach us for the hardships they must endure on our account. This is not to deny that we might be causally responsible for those hardships, only that we can be morally blamed for the effect they have on people's lives since, absent the world-with-hardships, *those people* would not have been born at all. This means that they cannot rightly harbour the relevant set of negative emotions toward us and therefore they should have nothing to forswear.

This, of course, is Parfit's non-identity problem expanded so as to encompass the phenomenon of forgiveness. Many attempts have been made to counter or resolve the problem. I will begin by endorsing the approach of a subset of these. The approach in question denies a key premise in the non-identity problem about what it means to harm, namely that to harm a person involves making that person worse off *than she would have been* had the harming agent acted otherwise.[2] This view is essentially comparative. 'Harmed' agents have received the benefit of existence, and non-existence is the only other option compatible with their not having been 'harmed' in these ways. But since it is better to be alive than not—that is, they have been benefited by being brought into existence—they have no moral case against their 'harmers'.

Now there is one rather obvious way to solve Parfit's puzzle, and it is important to mention it here in view of the specific threat we are talking about. Coming back to the seventeenth-century response to climate change, we find a plethora of anecdotal evidence indicating that for many people life was in fact not worth living. Consider this picture of life in Shandong, China around 1670:

> Many people held their lives to be of no value, for the area was so wasted and barren, the common people were so poor and had suffered so much, that essentially they knew none of the joys of being alive . . . Everyday one would hear that someone had hanged himself from a beam and killed himself. Others, at intervals, cut their throats or threw themselves into the river. (Quoted in Parker 2013, epigraph)

It should be noted that one does not need to kill oneself to judge that life is not worth living. It may be that suicide is simply too difficult. We cannot be complacent about this becoming the settled view of things in many parts of a future world devastated by the impacts of climate change. To the extent that the people making this judgement are aware of how their world got to be the way it is, they would, it goes without saying, likely be consumed by resentment of the perpetrators (us). Although their existence is dependent on our poor choices, they can blame us unproblematically. Since we have forced a thoroughly poisoned chalice on them there's no question of a benefit here. But let's leave aside this bleak possibility for the moment (I will return to it in Section 7.3 below in the context of a discussion of the judgement of unforgivability) and ask if, short of it, there might be a way for future generations to possess the appropriate reactive attitudes towards us. This brings us back to the comparative claim.

Within the approach I am endorsing there are two broad strategies for countering the comparative claim. The first is to say that agents can be harmed even if they are also benefited, or if the harms and benefits are inextricably packaged together. This might be the case if the sorts of harms people of the future undergo because of our actions constitute, as Cripps puts it, 'central, alienating costs' to them (2013, 17). The second denies that existence is, strictly, a benefit or interest rather than the condition for all other interests or benefits, as well as harms (Weinberg 2008). Absent the coherence of an appeal to the existence-benefit we have conferred on them, we in the present are not off the hook for the harms we cause to future people. Both types of argument eschew comparison between the agent's state and the alternative of her non-existence. Instead, they refocus our moral attention on the reality of the harms we cause, a reality obscured by the comparative judgement. These two versions of the approach await further refinement but even at this early stage of development they offer compelling support for our intuition that we are doing something gravely wrong to future people in breaking the climate.

If this is right, then feelings of resentment by future people towards us can in principle be justified. I think it is difficult to escape the comparative judgement altogether, however. It is implausible to suppose that a future agent could simply forget about the fact that her existence is, in fact, contingent on our reckless choices even if she embraces the general thrust of one or another version of the non-comparative view. This might make irresistible the comparison between her life-worth-living and her possible non-existence. In the grip of this thought her hostility is bound to wane, if not disappear altogether. But even while entertaining this diluted resentment towards us, the agent's thinking might undergo a subtle modification. Perhaps she relaxes the focus on herself and her personal deprivations, takes a look around at the damaged *world* she inhabits, and wonders how it is that we could have decided on policies that brought this world, rather than some clearly available alternative, into being. Fiona Woolard notes that philosophers sometimes take Parfit's non-identity problem to be a counter to an otherwise 'emotive' appeal to what we are doing to 'our children's children'. The non-identity problem instead asks us to take the cool, rational view (Woolard 2012, 678). But I think this is a mistaken view of what Parfit's own solution to the problem implies. There is every reason to think that the train of thought I have just been tracking gives rise to significantly powerful hostile emotions directed at us.

Parfit thinks that the truth he has discovered about personal identity should be considered morally irrelevant, and that we ought to proceed on the assumption that in breaking the future we are indeed doing something we should not be doing. Here is the principle he cites to support his case:

(A) It is bad if those who live are worse off than those who might have lived. (2010, 118)

This is a bad state of affairs, Parfit claims, 'even though it will be worse for no one' (2010, 118). The trick is to switch from judgements of wrongdoing based on personal grievance to ones based on comparisons with likely levels of welfare in merely possible worlds. Parfit adds that in view of its unfamiliarity, 'we have yet to explain why (A) should have any weight' (2010, 118). Well, why should it? One reason would be that, despite appearances to the contrary, it does support a reasonably held hostile judgement on the part of members of 'those who live' towards those who made their world. But what would be the basis of this judgement?

Strawson's discussion of the reactive attitudes is helpful here because he is very careful to draw a distinction between resentment and indignation. In particular, we must underline the unique 'vicariousness' of indignation. After a discussion of resentment, Strawson claims that:

> one who experiences the vicarious analogue of resentment is said to be indignant or disapproving, or morally indignant or disapproving. What we have here is, as it were, resentment on behalf of another, where one's own interest or dignity are not involved; and it is this impersonal or vicarious character of the attitude . . . which entitle it to the qualification of 'moral'. (2008, 15)

Further, emotions like indignation, 'rest on, or reflect . . . the demand for the manifestation of a reasonable degree of goodwill or regard, on the part of others, not simply towards oneself, but towards all those on whose behalf moral indignation may be felt, i.e., as we now think, towards all men' (2008, 16). Whereas resentment is the reaction to personal wrongdoing, indignation has this essentially impersonal and even universalizing character (one's 'own interest' is 'not involved'). Let's return to the identity problem with this distinction in hand. Because of the contingency of personal identity, people of the future whose world we are in the process of breaking will be able to resent us in only a muted way (if at all). If this were the end of the matter we would be left with a puzzle. On the assumption that the difficulty of forswearing resentment is in large measure a function of the intensity with which it is felt, it would be very easy for future people to forgive us for breaking their world. That, however, seems to let us off the hook far too easily. The puzzle is resolved by noting that their hostility towards us is likely to take the form of moral indignation rather than (or in addition to) resentment. If we break their world they will doubtless hate us, but the hatred might express vicarious identification with someone or something else. There are at least three options here.

The first option is that the others on whose behalf this feeling is expressed are those who, in Parfit's formulation, 'might have lived' were we less morally reckless. If the notion of proxy indignation on behalf of merely possible people sounds strained, the second option picks out those whose lives have been made literally intolerable, the future analogue of the people of seventeenth-century Shandong. Finally, we might parse 'all men' in the quotation above as referring to cosmopolitan or impersonal morality. In this case, what future people resent in our actions is our disrespect for or insufficiently enthusiastic support for that 'institution', as Bernard Williams refers to it (1985 , Chapter 10). I want to expand briefly on this third option. The danger in referring to morality as an institution is that this locution might obscure the extent to which the thing in question is identity-forming for a people. Although we are deeply attached to many of our specific institutions—the banking system, the nuclear family, the nation state, the modern corporation, our rigidly hierarchical military, the prison system, etc.—we can surely imagine going on without one or another of them, or with some radically altered version of them. But my larger purpose in this book has been to suggest that Enlightenment morality is not like this. If we lose it, we lose ourselves. If future people were disposed to think this way, contemplation of our generation's cavalier flirtation with climate disaster—and the fragmentation of humanity it was likely to entrain—might result in intense moral indignation towards us.

With respect to the non-identity problem it seems correct to say that we are in the grip of a deep-seated impasse, the point of contention being comparative and non-comparative understandings of harming.[3] I don't have an answer to this problem and I suspect there isn't one. But it's a fruitful tension because it sheds light on the phenomenon of intergenerational forgiveness. It seems plausible to suppose that victims of intergenerational wrongdoing will be shuttled between resentment and indignation in contemplating and working through their challenges. A diminished version of the former will dominate their thoughts and moods when thinking about their personal deprivations and those of loved ones; the latter will come to the fore when they abstract from these struggles and think about the world they inhabit compared to other possible worlds, the harsh existential struggles of many of their contemporaries, or the damaged institution of morality they have inherited. Since on this scenario a potentially powerful set of reactive attitudes is at play, logical space for forgiveness opens up. Nor do I see any reason to doubt that this phenomenon can take root in the minds of people whose negative attitudes are directed at an entire group. Even so, we might wonder what is so important about their retaining justified access to indignation. After all, the emotion is a painful one to harbor, so wouldn't they—or anyone—be better off without it, or by having less of it in their lives?

7.3 The Importance of Intergenerational Forgiveness

Strawson's discussion of the dangerous consequences of adopting an objective attitude towards other agents might help answer this question. The discussion is germane because, in denying that future people can rightly possess reactive attitudes directed at us, we are implying that they should view us generally in a more objective manner. All of this follows on a certain understanding of the non-identity problem. By being stripped of the opportunity to blame us for what we have bequeathed them, future people in effect deny our agency. But in a world struggling to cope with disaster it is a simple step from here to a wider denial of agency because members of any struggling generation might judge that they themselves will eventually be seen as non-agents—in the same way and for the same reasons—by people in subsequent generations. But what happens when a generation decides that its members are no longer agents in a robust sense? The answer is that they have made possible the 'thoroughgoing objectivity of attitude' Strawson was worried about:

> The only operative notions invoked in this picture are such as those of policy, treatment, control. But a thoroughgoing objectivity of attitude, excluding as it does the moral reactive attitudes, excludes at the same time essential elements in the concepts of moral condemnation and moral responsibility. This is the reason for the conceptual shock. The deeper emotional shock is a reaction, not simply to an inadequate conceptual analysis, but to the suggestion of a change in our world. (2008, 20)

What I'm interested in here is the possibility that by eliminating the judging gaze of future people we remove what could be the final moral constraint on our treatment of people of the present. Since by hypothesis we are already trying to cope with disaster, we might assume that virtually anything is permitted. In this case the tempting thought for our political elites is that we no longer have the luxury of viewing our contemporaries as agents, reducing them in effect to instruments of 'policy, treatment, control'. This is the negative corollary of an argument I made in Chapter 6 about the manner in which taking up the perspective of the far future can act as a moral constraint inspiring us, for instance, to guard the impartiality of our survival lottery.

Think of how this darker state of affairs might develop in an officially decreed state of emergency. An emergency happens when events have over-whelmed us. In genuine emergencies it makes some sense to say that our agency should be diminished because agency functions in the normal case by reviewing future options and choosing rationally among them. In an emergency, events happen too fast, are too unpredictable, and have conse-quences we could not have foreseen. The best thing to do in these circum-stances is to see to it that everyone is following a simple set of rules: evacuate the building, obey the curfew, put your own oxygen mask on before your child's, etc. But we also know from experience that it is all too easy for the authorities to act heavy-handedly in these situations. I argued in Chapter 3 that part of the problem with securitizing climate change is that it opens the door to narrowly military responses to the task of adapting to climate change.

The iterative logic of Gardiner's IGAR, examined in Chapter 6, illuminates this further. Being unprepared to meet its challenges with renewable energy, one generation burns large quantities of its stock of fossil fuels in order to cope with climate change–induced disasters, thereby worsening the climate for the next generation, which (unprepared in the same way) responds to its disasters by burning more of its stock of fossil fuels, and so on. This is the death-grip of perpetual emergency. It is all too easy for the authorities to say they *have no choice* but to act in a manner whose result is to propel the arms race ever forward. They would doubtless invoke the need to enhance security in doing so, but the important point is that in speaking and behaving this way they will have become the sort of practical determinists Strawson was so worried about. In the age of climate change, this is the sort of thing that looms if we accept one consequence of the non-identity problem, namely that no generation can rightly hate the previous one for the world they have inherited. If this claim is correct, then no generation should worry about how it will be seen by *subsequent* generations, a stance that will enable unspeakable behaviour in hard times.

All of this has a larger bearing on how we see ourselves historically. The threat for any generation in the midst of severe climatic disruption is that it might gradually lose the ability to see its work as part of a larger human story. This is because an essential feature of this sort of narrative-construction is that agents view themselves as morally responsible for the outcomes they help instantiate. But that responsibility is just what Strawson suggests is under threat by the universal adoption of the attitude of objectivity towards other agents. Think of how we view the notable achievements of past generations, like the abolition of the North Atlantic slave trade. They were essentially—though not exclusively—moral struggles, where individual agents as well as the collectives they made up had choices available to them and were responsible for having chosen one path rather than another. Appiah's description of the four moral revolutions, examined in Chapter 4, is instructive in this regard. In each case, the moral progressives fought hard to change offensive social practices. Eventually the rest of society came around, but nobody can really believe that in doing so agents were simply the playthings of extra-human forces (natural or divine), that they had literally no options.

If these things *were* true of humans as historical beings, there would be no point in telling a story that begins or continues with the struggles of past generations and ends or continues further with ours. Or rather, the story could not be *our* story as distinct from a story about us told by someone or something else.[4] The latter objectifies us, the former stresses our agency. Note, finally, that the danger to which I'm pointing is iterative and cumulative.

As each generation abandons the task of placing itself in a larger narrative framework the gaps in the story become ever larger, the counterexamples to it ever more ubiquitous, until the whole thing collapses into incoherence or becomes a joke. Justified access to the reactive attitudes that are integral to the process of forgiveness is as essential across the generations as within them because this is the only way we can retain a view of ourselves as *participants* in our own historical narrative. We require recourse both to indignation about our ancestors' moral failings and admiration for their moral victories. Absent such recourse the most we can say is that things happened to them, some of which turned out to benefit us while others did not. This sort of historical objectification confounds our intuitions.

Griswold is to be commended for having recognized that historicity is a key component of forgiveness: 'Forgiveness understood as a process, rather than simply as the end result, is much concerned with the temporalization of relations...and narrative is a form of explanation ideally suited...to articulate and convey a unity-making perspective through time in a way that attempts to make it meaningful' (2007, 187). If we work at it and are also lucky, people of the future—even those coping with the disasters of a 5°C global temperature anomaly—will tell stories about us. In the process, they will inevitably articulate their narrative in a way that helps them make sense of the world they live in and our role in bringing it into being. This means that we may at least be candidates for their forgiveness. But *will* they forgive us? Actually there are three distinct possibilities here. We will be judged forgivable (and then presumably forgiven), we will be judged unforgivable, or we will simply be forgotten and the question will never come up. I said we will be lucky if they tell stories about us in the future because this means that we have not been forgotten. The people discovered by Lambert in the former Scotland inhabit a world from which all record of our existence— those of us who fought for social and political justice no less than those who fought against it or were simply indifferent to it—has been expunged, and Wright's message about them is clear: this option represents the descent to barbarism.

Indeed, there are no coherent stories at all in that world beyond those required to make sense of immediate needs. I'm not sure there's much more to say about that option. If that is our lot, civilization may or may not recover, but I see no grounds for reasonable belief one way or the other. The options we need to spend more time examining are therefore the other two: the judgement that we are forgivable and the judgement that we are not. The place to start is with a critical look at the notion, most recently championed by Holmgren, that judgements of unforgivability are never justified. Holmgren argues that our response to wrongdoing must display a respect for the wrongdoer, for morality, and for the victim. The main claim is that a victim's retention of resentment or indignation—i.e., the refusal to forgive, perhaps consequent on a judgement of unforgivability—is always a mark of disrespect of the wrongdoer. Holmgren's analysis is useful here because of the way she invokes Strawsonian considerations:

> While it is reasonable to claim that the offender is responsible for his actions and attitudes, we ... objectify the offender in a way that is morally problematic if we conflate him with his actions and attitudes. In adopting an attitude of resentment, we may commit this moral error. We slide from the reasonable belief that the unrepentant offender is responsible for the offence and his current lack of remorse for it to the vague claim that he is 'identified' with them. (2014, 87)

For this reason resentment toward an offender and respect for her are incompatible. And since forgiveness is just the forswearing of emotions like resentment, the refusal to do so with respect to a particular offender is a mark of disrespect towards her. The conclusion is interesting precisely because it runs against the current of much philosophical thought about what it means to respect agents. On this understanding, which has its roots in Kant, blaming agents for wrongdoing—which often includes the holding of relevant reactive attitudes towards them—is precisely a way of holding them responsible and thus respecting their free agency.

Unfortunately, Holmgren's case rests almost entirely on the uncareful way in which the opposing view has been put by some philosophers. For example, Joanna North has argued that in cases of serious wrongdoing leading to judgments of unforgivability, 'the more we understand, the more we come to regard the wrongdoer as culpable, as wholly and utterly bad', and she says explicitly that in these cases we are right to identify the wrongdoer with her action (Quoted in Holmgren 2014, 86). I agree that this represents an objectification of the wrongdoer, but it is also, alas, not how we should think about all serious wrongdoing. Whether the account applies rightly to some wrongdoing, for instance that of the psychopath—in which case, what's the problem with objectification?—is beyond the scope of my analysis, but it clearly does not describe most of the wrongdoing I have been talking about in this book. I have argued that with respect to our failure to address the problem of climate change we are culpably morally weak and self-deceived. But this account presupposes that we are *not* identified with our morally objectionable actions. We are failing (a) to choose and act on values and principles whose relevance to our circumstances we recognize; and (b) to structure our beliefs fully in accordance with what we have good reason to believe is true.

So if there is no need to say that in doing serious wrong in this case we are to be identified with our actions, then the ground for the further claim that resentment of or indignation toward us for our failings is disrespectful has been removed. It follows that a judgement of unforgivability by future people towards us might be warranted. This brings us back to the suicide-ridden world to which I alluded in the previous section (7.2). Here, Holmgren's analysis alerts us to a problematic way judgments of unforgivability can play out. In *The Birth of Tragedy* Nietzsche describes a possible world in which people 'have learned to regard their existence as an injustice, and now prepare to avenge not only themselves, but all generations' (1992, 111). This outcome is scarcely more palatable than the one in which we are forgotten by the future. For example, we might be tempted to characterize those trapped in the awful, multigenerational swirl of the IGAR as dominated by this vengeful stance. They can neither forgive nor forget members of the generations preceding them. Clearly the judgement of unforgivability in this form is very harsh for both victim and wrongdoer. It sentences both parties—whole generations in this case—to a kind of moral shadow-existence, where everyone is psychologically simplified: dead generations becoming identified with their wrongdoing, the living with their own suffering. Can we avoid this outcome?

Sometimes, judgements of unforgivability follow simply on the severity of an offence. The offence is deemed by its victim to be so reprehensible that no amends-making by the wrongdoer can qualify her for forgiveness. That may be the judgement awaiting us, but it need not be. Another way to merit the judgement of unforgiveness looks to the inadequacy of the wrongdoer's response to the victim rather than (or in addition to) the wrong itself. On this model, if the wrongdoer, aware of the wrong, neither takes steps to change her behaviour nor engages in any subsequent morally reparative acts vis-à-vis the victim, she may also be judged unforgivable. This is especially relevant for the intergenerational case, since we will all soon be dead and therefore unable to change our minds at some point about the advisability of amends-making. What we do, or fail to do, will become a part of our permanent moral record. Let's assume for the sake of argument that it is at least possible for us to escape the judgement of unforgivability by future generations for the havoc we are bringing to their world. That is, that in spite of the harms we are causing there is something we might do to avoid this judgement and instead be forgiven. What should we do? I don't mean this question to be very specific, because I have already answered it at that level: we should decarbonize the global economy rapidly. What I am asking now is how we should understand this task in a way that exposes us to the moral gaze of future people and thus expresses our common bond of identity with them.

Griswold has identified six conditions an agent seeking the forgiveness of her victim must fulfill, and his list can be adapted to our problem. He argues that the wrongdoer must: (1) take responsibility for her deed; (2) repudiate the deed; (3) express regret at the injury caused; (4) be committed to becoming a better person; (5) sympathize with the victim; and (6) offer a narrative account of her wrongdoing (2007, 49–51). The question for us to ponder is what actions we in the present can take to make it the case that we have fulfilled these conditions even as we are in the process of inflicting harms on future generations through our profligate GHG emissions. There are probably options available to us with respect to conditions (2) to (5), though it is difficult to see what repudiation of the deed (in condition (2)) could mean in this case if it does not include rapid decarbonization of the global economy. Apart from this, it seems to me that the sincerity of our efforts with respect to these conditions will be judged mainly on the basis of how we deal with conditions (1) and (6).

Condition (1) is clearly important for it requires us to have recognized that we are morally blameworthy on account of our current practices and that we need to be honest and precise about the scope of the harms we are causing. Much of my argument in this book is designed to show that we are very far from meeting this condition and thus accepting full responsibility for what we are doing. Chapter 3 provided an examination of an important subset of the injuries we are causing, while Chapters 4 and 5 were meant to diagnose the various ways in which we have attempted to evade full responsibility for what we are doing. I won't repeat the arguments here, only point out that until we resolve the deep-seated moral-psychological roots of our crisis, our efforts to meet conditions (2) to (5) are likely to be hollow and self-serving. Indeed, I would suggest that this is exactly how we should interpret a good deal of current liberal hand-wringing about climate change. And we should also be worried about the comfort provided by a certain way of understanding Parfit's non-identity problem. When teaching climate ethics I am always dismayed to see how warmly my most reactionary students greet the *problem*, while eagerly dismissing Parfit's 'solution' to it.

In any case, a better way of going about making real changes may be to start with a version of Griswold's condition (6). I say 'a version of' this condition because although this book has been focused on the nature of the wrong that is climate change, we also need to emphasize the positive aspects of the narrative I have been constructing if we want to motivate people to make the sweeping changes to their lives that are now necessary. This is why I began by defining the Anthropocene Project in a way that stresses its connection to values and principles many of us already endorse. These days we are often told to think small, move more slowly, act locally, pursue incremental change at sub-national levels, and so on. There's something important about these injunc-tions, but perhaps our inertia can be explained in part by a nagging sense that we're not entirely sure what their larger point might be. This thought can become especially troubling when we emphasize, as I have, the need for sig-nificant struggle and sacrifice. I have focused unabashedly on the much bigger historical picture, a perspective I think has been sorely missing in our approach to the problem so far. One inevitably loses sight of a few details in an exercise of this sort, especially on the policy front, but there are plenty of intelligent people—both across the academic disciplines and beyond the academy—working out such details, so I don't see this as a problem for my account.

The phenomenon of climate change should be situated in the context of a conscious attempt on the part of our species to give a specific character to our social and historical existence. It concerns how to live more justly and intelligently together, and we have been constructing it, at least in a concerted and systematic way, for at least 250 years (though some of its philosophical roots are much older). Because it forces our moral gaze so far forward in time, climate change poses new challenges to this enterprise, and that's why the Enlightenment Project has become the Anthropocene Project. The Enlightenment saw itself historically too, as I have emphasized in Chapter 2. But its future gaze was rosier than ours can be. Unvarnished Enlightenment optimism has been replaced by an acute awareness of moral danger, of the harms we are capable of visiting on people of the future as much as the benefits we can provide them. As I have said, this is the upshot of the way in which our - causal—that is, technological—power is now combined inextricably with imperfect control over its products. But the animating spirit of the two historical and moral projects is the same, and it is that spirit I have tried to give shape to in these pages.

7.4 Conclusion

In the end I doubt we will get very far in our current predicament unless and until we place it in the context of this specific and larger historical narrative. The precise nature and full extent of our wrongdoing come into view only if we begin by understanding ourselves as the inheritors of a moral tradition committed to the centrality of the virtues of justice, truthfulness, and rational hope in the moral life. If we can internalize this thought fully, we will see not only that we occupy a unique place in history but also that by dumping so much carbon into an over-stressed atmosphere we are in danger of undermining this extraordinarily valuable but fragile and still unfolding ideal. Embracing the Anthropocene Project in a way that avoids both hubris and timidity can act as a catalyst for us, inducing us to take full responsibility for our wrongdoing which, in turn, might lead us to take meaningful steps towards securing a decent future for humanity. We cannot force everyone to take the necessary steps, but all of us have the option of acting with integrity in this crisis. Although we—the global prosperous—cannot avoid injuring them to some degree, if enough of us can show that our hands are clean we will have given people of the future a reason to forswear their justified moral indignation towards the lot of us. If they do, it will mean they have found a way to continue our story.

Notes

Chapter 1

1. As Goldstone notes, the Netherlands, which experienced its 'Golden Age' at the very same time as the Little Ice Age, provides an interesting counterexample to Parker's thesis about Japan's uniqueness in this period. Goldstone (2013, 37).
2. See also Moellendorf (2014).
3. One way to be more precise about this feature of the global prosperous is to cite AR5's classification, based on World Bank data, of countries into 'income groups'. In this case, the global prosperous would be people living in 'upper middle income' and 'high income' countries which were responsible, respectively, for 18.3Gt and 18.7Gt of CO_2 emissions in 2010. Of course, these data hide wealth, and therefore emissions, disparities within these countries and this must also be taken into account. And there will be some countries in these brackets whose citizens for the most part do not fit features (2) and/or (3). But the classification is nevertheless helpful. See IPCC (2014b, 14–17).
4. The point made here by both Jamieson and Sandler is directed at utilitarianism alone (although it might extend to certain forms of contractualism as well) and does not therefore undermine a deontological approach to collective action problems. Deontology has its own problems but the contingency issue is not one of them.
5. Slote distinguishes between agent-focusing and 'agent-basing'. Agent-based virtue ethics, such as the one advocated by Slote himself, takes character as explanatorily primary. Actions and the consequences they help to bring about are right to the extent that they are the product of virtuous dispositions. I think that agent-based virtue ethics is too radical. Absent a strong connection between the exercise of the virtues and the goal of producing morally right consequences, Slote ultimately fails to say enough about why we should care about cultivating one set of virtues rather than some other. However, this is a large issue in normative ethics and I don't pretend to have made the case against agent-basing here. See Slote (2001) and (1997) for further elaboration of these themes.
6. The deepest accounts of agent-focusing, I think, are those of neo-Aristotelians like Hursthouse, Foot, and Annas. Much of what I say throughout the book is inspired by this general approach, though there are of course important differences among these philosophers (and other neo-Aristotelians) which I do not address. See Hursthouse (1999); Foot (2001); Annas (2011).

7. Ronald Sandler has argued persuasively that scientific explanations of morality fail because (a) science cannot make full sense of beliefs, desires, and actions, though these are key objects of our evaluative practices; and (b) science cannot fully specify what our ends should be. See Sandler (2007, 19–21).

8. See Tuck (2009) for a thorough airing of these issues. Melissa Lane has a lengthy discussion of the negligibility thesis in which she claims that it rests on the false assumption that because climate change is a big problem it must have a single big cause, and the trick is to find that one big cause. The truth however, as Lane sees it, is that it is the product of millions of micro-decisions. Because of this there is no threshold for making a literal difference to the problem: 'every single atom of greenhouse gas emissions saved is a literal improvement' (2012, 59). Further '...it is illogical to take refuge in being oneself only a small percentage of the problem because there is no other comparable body whose contribution is of a dramatically different order of magnitude and who therefore has significantly more reason to act than oneself' (2012, 59–60). I fail to see how these claims resolve the problem. While it is true that every carbon molecule not emitted lessens the total amount in the atmosphere, compared to the counterfactual state of affairs in which that amount *is* emitted, the foregone amount might be small enough to be inconsequential in terms of its effects on global climate. So the corrosive and reactive reasoning that defines the negligibility thesis remains in place. Nor do I see what is 'illogical' in the claim that there are larger entities than me at play here—whole countries, for example—whose legislative actions can have a much bigger impact on the problem than anything I might do. So these 'direct effects' of my actions on the climate do not make the negligibility problem go away. Lane also discusses 'indirect effects' my actions might have on others, like a neighbour being inspired to do more for the environment when she sees me putting solar panels on my roof. Whether these effects will be substantial, and therefore affect the negligibility thesis, is an entirely contingent question.

9. Kant's way of thinking about freedom is relevant here because he insists on a distinction between the practical and theoretical standpoints. From the former standpoint, that of the phenomena, the world is fully deterministic; but the practical standpoint, expressed in the adoption of moral principles is, for Kant, 'a dimly conceived metaphysics'. That is, when we act morally—by constraining this or that desire, for example—we must suppose we are free even if we cannot explain our freedom from the standpoint of theoretical reason. See Kant (1983, 376). Since intergenerational ethics involves constraining our generation-relative impulses, the point I am making here can be viewed as an application of Kant's general principle. Evolutionary psychology, sociobiology, and game theory—to the extent that each of them in its own way confines our practical judgements to the calculation of our own gain, or that of those genetically related to us—cannot make sense of our agency as applied to this problem. It is moreover question-begging to claim that we cannot as a matter of fact transcend our own perspective in the required fashion. We plainly can. How else do we distinguish plausibly between (a) the person who constrains his consumption out of some neurotic fixation on

his own undeservingness or in order to gain a reputation as an environmentalist; and (b) the person who does the same thing because he judges that future people have a moral claim to some of his resources or because he empathizes with their likely deprivations? Since they will inevitably reduce motives of type (b) to those of type (a) the perspectives I am criticizing cannot make sense of our practical lives.

10. Pascal Bruckner thinks that Europe is currently awash in guilt and that this is hindering it from approaching its cultural tasks boldly, the way America allegedly does. There is no doubt that this sort of attitude can become pathological, but Bruckner surely overstates the extent to which it has done so in Europe. He also fails to see how the kind of historical oblivion he thinks is laudable is preventing Americans in particular from recognizing their disproportionate complicity for the broken climate. I address this question more fully in Chapter 5.4. See Bruckner (2010).

11. The following three paragraphs are based on Williston (2011).

12. See Heath (2014) who adapts Andy Clarke's notion of the mind's external scaffolding in this way.

13. A complaint originally raised about modern morality by G.E.M. Anscombe. See Anscombe (1958).

Chapter 2

1. It might be fruitful to situate this claim in the context of Jamieson's recent analysis of the issue. Jamieson distinguishes between recognizing our moral commitments, extending them, and revising them. So, for example, Peter Singer's arguments about animal liberation call on us to *extend* our moral principles, while his arguments about famine call on us to *recognize* the moral commitments we already have. On the other hand, Jamieson thinks that most climate ethicists are asking, even if only implicitly, for a wholesale *revision* of our moral concepts, even, possibly, to the point of advocating for 'moral revolution'. In my analysis, these categories are not so distinct, but I'm not sure there's a problem with this. For example, in this chapter I show that we have reason to extend cosmopolitan principles most of us already espouse to people of the future. In Chapters 4 and 5 I argue that we are not fully living up to—that is, recognizing—our own best moral ideals. And in Chapter 6 I suggest that engaging in the requisite extensions and recognitions would likely result in a sweeping revision of our social and political practices, if not our moral concepts. See Jamieson (2014, Section 5.6).

2. For the record, I do not count myself an advocate of a 'Good Anthropocene', at least as this notion has come to be understood in the public pronouncements of self-styled 'eco-pragmatists' like Ted Nordhaus, Emma Marris, Erle Ellis, Stewart Brand, Mark Lynas, Andrew Revkin, and others. On this view, we should embrace optimistically our new role as planetary masters because our technology and our ingenuity are bound to save us. Though many of them would no doubt resist the claim, eco-pragmatic techno-optimism plays directly into the hands of cruder accelerators like Tillerson. Clive Hamilton's analysis of this group is appropriately

withering: '[i]t is not surprising that the eco-pragmatists attract support from conservatives who have doggedly resisted all measures to cut greenhouse gas emissions, defended the interests of fossil fuel corporations, and in some cases worked hard to trash climate science . . . [T]he "good Anthropocene" is a story about the world that could have been written by the powerful interests that have got us into this mess and who are fighting so effectively to prevent us from getting out of it'. Hamilton overstates his case, however, in deriding what we might call the group's characteristic opportunism (as distinct from its optimism). Since our power to affect the earth is now undeniable, the eco-pragmatists are surely right to frame our consciousness of the Anthropocene as an opportunity for change. After all, the only alternative to this is a form of fatalism. See Hamilton (2014).

3. On racism, see Hume (1882). Kant (2011). On sexism, see Kant (2011); for a careful treatment of Rousseau's views on women, see Shell (2001). It is crucial to note, however, that on these important issues, there were plenty of countervailing voices, voices that spoke from within the conceptual terrain of the Enlightenment itself. On anti-racism, see Woodman (1995). On anti-sexism, see de Gouges (2013).

4. They are incomparable, Harris claims, because of the falsity of the 'trichotomy thesis'. According to this thesis, there are (at least) three terms in any comparison, the two things being compared and a covering value that allows for the comparison. So the comparison between a salad and a hamburger as my lunch options can be made by reference to the covering values of cost, my health, what is better for the planet, etc. Harris argues that 'even under favorable epistemic and psychological conditions, rational people cannot always compare the value of options in a way that allows them to choose'. In fact, as the rest of the argument makes clear, this is so for virtually all important comparisons. This would mean that we cannot make diachronic value comparisons of the sort crucial to my arguments in this chapter. After all, such comparisons depend on the truth of the trichotomy thesis: we want to be able to say, for example, that liberal-democratic judicial institutions are better than their premodern counterparts with respect to the covering value of promoting legal equality. See Harris (2006, 50–1).

5. To be precise, Harris thinks that we can understand the idea of moral progress at two distinct levels. The first is the personal. Here the belief in moral progress is irrational and can be sustained only if the fact of incomparability is pushed out of consciousness. The second is the global. In this case, Harris endorses the possibility of progress in science only, and in particular of scientific psychology. The latter will increasingly allow us to understand 'our values and the conflicts among them'. But not only is this not what we mean by moral progress, such knowledge might actually make such progress more difficult by heightening our appreciation of the (alleged) incomparability of our values. Harris (2006, 270).

6. Satisficing is the procedure of (a) setting 'minimum aspiration levels' for desired outcomes (baselines below which you will not go); then (b) choosing the option that is a 'good enough' instantiation of your desires above the baseline. Given inevitable time constraints on our choices, it is meant specifically as a way of avoiding the decisional paralysis of optimal or maximizing approaches to deliberation. See Simon (1955).

7. Harris is therefore incorrect to see pessimism as the most potent threat to his version of tragic pluralism. Nevertheless, he does claim to overcome even pessimism. But the only way he can do this is by abandoning, without defeating, the impersonal standpoint that generates the pessimistic challenge, and adopt instead a kind of optimistic personalist perfectionism. How he can both admit that this does nothing to defeat pessimism on its own terms and insist that we should live without 'pernicious fantasies' is not explained.

8. The cuts required are very steep, and we should not be misled by tokenism. Although it goes beyond mere tokenism, the deal signed between the U.S. and China in November, 2014 is not nearly as significant as some tout it as being. The deal would see the U.S. cut emissions 26%–28% below 2005 levels by 2028, while China's emissions would peak around 2030. Not only does this scheme lock in current emissions rates for at least a decade for both countries (and longer for China), it still involves our exceeding the carbon budget by 2042. See Commodities Now (13 November 2014).

9. See McKinnon (2011 Chapter 5).

10. For example, why separate airline corporations from mass-market capitalism as features of, respectively, levels two and three? Since the former cannot exist in their current form absent the latter they must belong to a single level of analysis. To my thinking, the best way to separate levels two and three—though I admit that the line between them is probably necessarily blurry—is to confine the former to large-scale systems while the latter adds details about (a) the way geopolitical forces shape those systems; and (b) the way individuals internalize the norms of those systems. This is what I try to do in my application of the model to climate change.

11. According to Ruddiman, there is evidence that agricultural practices, especially rice cultivation, have influenced global climate since the dawn of the agricultural revolution. Also salient from this perspective are plagues which, by wiping out huge numbers of humans, allowed for significant reforestation. Ruddiman (2005).

12. For the role of climate change in these extinction events, see Ward (2007, 166–99).

13. For the juggernaut metaphor, see Giddens (1990, 136).

14. For a general defence of this view of progress, see Godlovitch (1998, 282).

15. The possible harms differ depending on the technology. One of the most popular schemes involves loading the stratosphere with sulphur dioxide which when combined with water vapour forms sulfuric acid. This reflects sunlight and can therefore cool the planet. But adopting this method does nothing to control the amount of carbon we pump into the atmosphere. So even if we can keep the planet cool, we might end up increasingly acidifying the oceans, with results that would likely be catastrophic for marine life (and therefore also for us). This example illustrates the overarching problem of moral hazard that besets all geoengineering ideas. If we believe that our technology is protecting us, we will likely engage in riskier behaviour. Again, I don't think risk can be eliminated in the Anthropocene, but we should not allow its parameters to be *defined by* the technological imperative. In my view this is what is currently happening with respect to geoengineering.

16. A good survey of the ethical issues of solar radiation management can be found in Preston (2013).

17. One suggestion as to how we can avoid depriving the poor of access to life-sustaining fossil fuels under a cap-and-trade system is to divide emissions permits (under the cap) into those which are free and those which must be paid for, with the former going entirely to the poor. See Shue (2014, Chapter 16).

18. Miller (2010, 382).

19. The claim is supported by three considerations, only the third of which strikes me as being potentially compelling. The first two are as follows. First, 'relations among compatriots' are intrinsically worth having. The claim here is that although national membership is instrumentally valuable, such value can exist only on the bedrock of intrinsic value. The evidence for this claim seems simply to be that this is in fact the way people view national membership. The second feature is that 'special duties to compatriots are integral to the idea of nationhood'. This is not to be confused with the 'purely cultural' way of looking at nationhood. According to the more robust conception of special duties Miller has in mind, while I've doubtless got ties to other members of my nation based on our mutual attachment to certain cultural achievements (like fans of a certain celebrity), I have no moral responsibilities to them beyond this. By contrast, for Miller our sense of national identity is the basis of deep political values like 'social justice or deliberative democracy'. But, of course, it might be the source of indecent values too, which is why everything hangs on the third consideration, analysed in the next paragraph (Miller 2010, 385).

20. See also O'Neill (1996, 174–8).

21. This is the key phrase of the original 1992 United Nations Framework Convention on Climate Change (UNFCCC). For a succinct analysis of what it means, see Mann (2009).

22. Which he defines as follows: 'not only will it be the case that every emission produced by one source is an emission that may not be produced by another, but some of the emissions permitted in a given year will have to be forgone by everybody in the subsequent year' (Shue 2014, 99).

23. See also Appiah (2010).

24. A notable exception to the tendency to overlook Enlightenment conceptions of virtue is Griswold (1998).

Chapter 3

1. As AR5 makes clear, the temperature anomaly range (in 2100) for RCP 8.5 is 2.8°C–7.8°C, with a likely outcome of 4.1°C–4.8°C. For simplicity's sake, throughout the rest of this book, I will refer to this as the 5°C outcome. See IPCC (2014b, Table TS.1, 26).

2. Corroboration of these claims can be found in Solokov et al. (2009); World Bank (2013), 7–18; Riahi et al. (2013); Schaeffer et al. (2013) (the latter two sources are cited in World Bank [2013]).

3. The emissions deal signed by China and the U.S. in November 2014 does little to change this, and certainly does not, by itself, warrant abandoning RCP 8.5. See note 8, Chapter 2 for details.

4. For a summary of all these faulty predictions, see Romm (31 January, 2013).

5. My discussion here differs from the way this issue is usually analysed. For example, Roger Pielke Jr has criticized climate scientists for being politically motivated in too many of their public pronouncements, arguing in effect that this stance has distorted their science. Jamieson thoroughly debunks Pielke's arguments in support of this claim, so I won't pursue the matter myself. See Jamieson (2014, 68–70).

6. On the need to engage the two information-processing systems, see Weber (2006). For an argument that positive emotions and aspirational values are important in the communication of risk and danger, see Public Interest Research Centre (2013). The idea that overexposure to catastrophe induces psychic numbing can be found in Slovic (2007, 2013). For related discussions, see O'Neill, Saffron and Cole (2009); Brulle, Carmichael and Jenkins (2012); Gardiner (2011, 193, n. 27).

7. The best discussion of this issue is in Jamieson (2014, 188–93).

8. The other tension belt is in high-latitude regions of the north. Here, the flashpoint for conflict could be the relative abundance of resources, like oil and gas beneath the Arctic seabed, being made available by a warming climate.

9. It is crucial to mark the distinction between climate refugees and climate migrants. Migrants are very often 'pulled' to the place to which they migrate, for instance by the prospect of greater economic opportunity than was offered in the home country. Migration is a recurrent feature of the movements of humans across the globe and can be entirely benign. Refugees, by contrast, are 'pushed' out of their homeland, either by adverse environmental or economic conditions at home or by forced expulsion at the hands of the government or some other ethnic group. But since the lands to which they are moving are not necessarily places of resource abundance and economic opportunity, the new arrivals may be perceived as threats to the vital interests of those who are already there. Of course, people on the move can believe they are being both pushed and pulled out of the homeland and, to the extent this is the case, the results—in terms of the potential for conflict—will themselves probably be mixed. See Homer-Dixon (1994).

10. A good discussion of this material as it relates to the connection between climate change and genocide is in Alvarez (2009, 131).

11. Of course, this is not to suggest either that the developed world should not be pursuing adaptation measures or that the developing world should not be thinking about mitigation. In the former case, our preparedness varies across jurisdictions but is generally not nearly as good as it should be; in the latter case, the goal of global decarbonization includes funding by the developed world of green development in the developing world (as I have argued in Chapter 2). The present point is simply that the short-term *priority* of the developed world is to mitigate net global emissions while that of the developing world is to look for ways to adapt intelligently and justly to the negative effects of climate change that are in the pipeline.

12. See especially Caney (2010); Hiskes (2008).

13. Appeals to rights can also be powerful motivators, as those who have fought for equality over the years and in various contexts can surely attest.

Chapter 4

1. See Brown (2013); McKinnon (2011); Posner and Weisbach (2010); Shue (2014); Caney (2010). A notable exception is Lane (2012).
2. See Aristotle (1947, Book VII).
3. To be precise we are currently in what Gardiner refers to as one of the 'degenerate forms' of the PIP, where the provision of front-loaded goods by present generations is heavily favoured over that of back-loaded goods, even though the latter are not entirely neglected. Still, it makes sense to refer simply to the PIP in analysing the problem. As Gardiner puts it, 'cases with structures close to the PIP are likely to arise in practice, so that the PIP may be a serious problem in the real world'. See Gardiner (2011, 167–9).
4. Gardiner (2011, 154). This paragraph is a modification of Williston (2012a, 172).
5. A caveat is in order here. For the Greeks, *pleonexia* involves the willingness to engage in extreme measures in order to achieve one's purposes. Their analysis of this motivation is of a piece with the general attempt to come to grips with the problem of political tyranny. In linking *pleonexia* to modern consumption, Lane is somewhat uncareful about this, which is why I prefer to couch the discussion in terms of ordinary greed and its contrast with gluttonous desire.
6. A good historical overview is in Hoffmann (2008).
7. A good discussion is in Mele (2001).
8. The phrase is Aristotle's, used to describe the Socratic position. Aristotle writes: 'Now we may ask how a man who judges rightly can behave incontinently. That he should behave so when he has knowledge, some say is impossible; for it would be strange—so Socrates thought—if when knowledge was in a man something else could master it and drag it about like a slave'. Aristotle (1947, Book VII 1145–6).
9. There is a long-standing problem with appeals to integrity, namely that they invite the possibility of 'moral danger'. Obviously, an agent can be both 'whole' and immoral. For a thorough and effective defence of the ideal of integrity against this challenge, see Sherkoske (2012).
10. Woodruff (2011, 165–7); see also Korsgaard (2009).
11. To be clear, Socrates is not describing the mechanics of weakness here, so there is no contradiction between the view that weakness is impossible and the view, expressed in this quote, that we can fail at self-mastery. To be self-defeated or licentious is simply to allow vicious desires to overwhelm reason. This can happen only if the agent is *ignorant* of the good.
12. For the sake of simplicity I have been talking about our thinking of ourselves as forming a collective with people of the future. That is not, strictly speaking, the way Cripps understands the key concept. For her, a 'should-be' collective is a set of individuals fulfilling two conditions. First, they would constitute a collectivity if they were to espouse certain goals together; second, they have a duty to espouse those goals. Cripps (2013, 60). Cripps is talking about the importance of forming a collective of people of the present so that we might act in concert to solve the problem of climate change. But if we manage this, we will by that fact have formed a bond of identity with people of the future as well. This is because what we have a

duty to do is agree on genuinely sustainable policies, an agreement that would protect the vital interests of posterity. So it seems as though the really important collective is indeed the one comprising all present people as well as people of the future, even if it is difficult to understand how the latter might 'espouse' the requisite goals. Still, I'm inclined to think it legitimate to speak as though they would do so since no matter who they turn out to be they will presumably prefer to inherit a non-degraded planet. This raises large issues, however, not the least of which is Parfit's non-identity problem. I address this issue, albeit with a different argumentative focus, in Section 7.2, below.

13. See Shue (2014, *passim*) for extended discussion of what constitutes fair shares in the climate case.

14. My use of the term 'luxury' here is inspired by the influential distinction between luxury and survival (or 'subsistence') emissions first put forward by Agarwal and Narain. The distinction was made to combat a report put out by the Washington-based World Resources Institute (WRI) arguing that developing and developed countries bear an equal burden for cleaning up climate change. Agarwal and Narain see this claim, one that is still very much alive in public debates about climate change cost sharing, as an expression of environmental colonialism. They write: 'Can we really equate the carbon dioxide contributions of gas guzzling automobiles in Europe and North America or, for that matter, anywhere in the Third World with the methane emissions of draught cattle and rice fields of subsistence farmers in West Bengal or Thailand? Do these people not have a right to live? But no effort has been made in WRI's report to separate out the "survival emissions" of the poor, from the "luxury emissions" of the rich. Just what kind of politics or morality is this which masquerades in the name of "one worldism" and "high minded internationalism"?' (Agarwal and Narain 2003, 3); see also Shue (2014, Chapter 2).

15. We should not be complacent about the IEA's March 2015 report that because the global economy grew by 3% in 2014 while carbon emissions remained steady, the world effectively achieved decoupling in 2014 (without the 'benefit' of an economic recession). First, the result may be temporary. There is some reason to believe that it has much to do with China's success at harnessing hydro-electric power that year, and that this may not be a repeatable achievement. More importantly, as I have been arguing, we need carbon emissions to come down rapidly, not just plateau. Think of this in terms of the goal of carbon intensity reduction (another way to think about decarbonization, decoupling, etc.). While global carbon intensity fell by an average of 0.9% per year between 2000 and 2013, we need it to fall by *6.2% per year every year* until 2100 to avoid climate catastrophe. See PricewaterhouseCoopers (2014), 2-3. For further analysis of the 'myth' of decoupling, see Jackson (2011), Chapter 5. I thank an anonymous reviewer at OUP for suggesting I emphasize these skeptical points about decoupling.

16. 'Half measure' is probably better than 'non-solution' here. As Shue has argued, there are huge 'no-regrets' efficiency measures available to the rich. But he also recognizes that this approach will not get us all the way: they provide 'only a kind of grace period before the crunch comes' (Shue 2014, 137).

17. On rooted cosmopolitanism see Appiah (2005, Chapter 6).
18. The following two paragraphs are based on Williston (2012b).
19. See Velleman (2006); see also Williams (1993); and Taylor (1985).
20. Solomon is thus incorrect, in my view, to claim that shame necessarily involves taking responsibility for doing something wrong. See Solomon (2007, 60).
21. That is, agents in moral dilemmas are best conceptualized as blameless wrongdoers. There is nothing amiss in the way they deliberate and choose—hence shame is inappropriate—though their actions are harm-causing, which makes guilt appropriate. See Williston (2006).
22. See Woodruff (2011); and Cunningham (2013).
23. Two recent books by former oil executives provide illustrations of this phenomenon. The first is Hoffmeister (2010). The second is George (2012). In a manner that patently panders to the interests of the oil industry, both Hoffmeister and George argue that it is morally imperative for us to continue to access as much of the world's remaining stock of fossil fuels as we possibly can. The chief impediments to this crusade, especially for Hoffmeister, are government regulations of industry and the environmental interventions of the 'left-wing' judiciary.
24. For a critical analysis of these ideas as they developed historically, see White (1974).
25. See Simon (1998).

Chapter 5

1. Lomborg thinks there are much more urgent global public health issues we should be addressing in the present and that GHG abatement would therefore constitute a misallocation of resources. Since we are talking in this chapter about epistemic virtues, and their lack, see also Howard Friel's devastating analysis of Lomborg's shoddy scholarship, especially the latter's very suspect deployment of important data (Friel 2011). In my view, Lomborg has failed to provide a convincing response to Friel's accusations.
2. Shue points out, correctly, that the 'fatal flaw' in these arguments is that those likely to have the biggest challenges of adaptation (Bangladesh, sub-Saharan Africa, etc.) will also be least able to meet the costs of adaptation. At the very least, therefore, proposals like those of Lomborg and Beckerman require supplementation by consideration of how to allocate adaptation costs fairly. See Shue (2014, 139).
3. We hear about such intimidation most often in the American case, but there is perhaps no more brazen example of it than in Canada under the rule of Stephen Harper's Conservative party. In Canada, all government scientists—especially those whose work touches on the phenomenon of climate change—are now prevented from communicating their findings to the broader public. Instead, interested members of the public must seek information through a communications officer run directly out of the Prime Minister's Office, and this person invariably responds with the usual bromides about, for instance, the government's commitment to 'balancing' the economic and environmental concerns of Canadians. As is widely understood, the primary economic objective of the government is to ramp up production of the tar sands in Alberta, but Canadian citizens worried

about the environmental impacts of this development now have very little access to sound scientific data about it. An engaging and well-researched report about all of this is in Turner (2013).

4. Vanderheiden provides a thorough analysis of the issue (2008, Chapter 1).

5. See McKibben (2012, 164).

6. See Roberts and Wood (2007, 157–60). Roberts and Wood analyse several other criteria as well, but I am less convinced of their importance.

7. See Volk (2008) for a splendidly detailed—and highly personalized!—description of the carbon cycle.

8. The example Roberts and Wood provide is that of the human genome (Roberts and Wood 2007, 158).

9. See Gribbin (2009, 4).

10. We must also note what sort of virtue epistemology this analysis seems to require. There is a traditional distinction in virtue epistemology between reliabilists and responsibilists. Reliabilists like Goldman, Greco, and Sosa understand the intellectual virtues to encompass processes and faculties. Thus memory, intuition, and perception are virtues. Responsibilists like Baehr, Code, Hookway, Montmarquet, Roberts and Wood, Sherkoske, and Zagzebski, by contrast, stress agency. Thinkers in this group, as Greco puts it, tend to be more concerned with 'the active nature of the knower, as well as the element of choice involved in the knower's activity'. This stress on activity and choice suggests strongly that we should understand climate change denial on broadly responsibilist lines. The same applies to excellent epistemic achievement. In this section I have described the ideal epistemic expert— represented by Lovelock and Leopold—while in the next section (5.3) I describe the ideal epistemic non-expert. Both are worthy of our admiration, a stance that seems to require responsibilism. Greco (2011).

11. For example, when Texas Congressman Steve Stockman (R) claims that climate change–induced sea rise cannot happen because *all* melting ice is like the proverbial ice cube in a glass of water, he enables just this sort of epistemic trickle-down. This kind of thoughtlessness, ubiquitous among American politicians, provides comfort to and cover for the intellectually challenged and the lazy. For a video transcript of Stockman's sorry performance in committee, see Atkin (18 September 2014).

12. At least this is a necessary condition for the justification of beliefs on broadly internalist grounds. For the internalist, beliefs can be justified only by reference to internal states cognitively available to the agent herself (other beliefs). Externalism, by contrast, is the view that the right sort of causal process leading to a belief can justify that belief, whether or not the agent is aware of the details of the process. For the many forms the two positions can take, see the excellent essays in Kornblith (2001).

13. Since the object of knowledge here is irreducibly social and highly refracted we would do well to think of the problem in relation to the three species of social epistemology outlined by Goldman. First, there are Individual Doxastic Agents with Social Evidence (IDA). Although it focuses on the epistemic state of individuals, 'it addresses doxastic choices made in the light of social evidence' Goldman (2011, 14).

Under this heading we find the problem of peer disagreement, for example. Second there are Collective Doxastic Agents (CDAs). Here the collectivity makes judgements about the truth value of propositions. Because its reports identify particular claims as 'likely', 'very likely', and so on, the IPCC is said to be a prime example of a CDA. Finally, there is Systems-Oriented Social Epistemology (SYSOR), examples of which include science, education, legal adjudication, and journalism. Goldman (2011, 17).

Two points about this classification. First, each species of social epistemology has its own distinctive concerns and questions. Thus IDA has as its focus traditional epistemological concerns with truth and justification from the standpoint of the individual doxastic agent, but as noted, the twist is that such agents are confronted with evidence that is inherently social in nature—communication by other doxastic agents, information from media, etc. CDAs are concerned above all with the problem of judgement aggregation. To what extent is the body responsive to the judgements of its individual members? Can we understand individual and collective rationality the same way? And so on. As for SYSOR, it attempts to ferret questions of justified belief out of the workings of 'formal institutions with publically specified aims, rules and procedures' (Goldman 2011, 19). The second point is that in spite of these special concerns and questions, there can be overlap among the forms. As Goldman notes, CDAs usually affect the beliefs of IDAs, and some entities can qualify as both CDAs and SYSORs. This is especially pertinent to our assessment of the IPCC. Goldman is surely correct to count the latter as a CDA, but the knowledge purveyed to the rest of the culture by this collective is more complicated. Since the IPCC reports—'Summaries for Policymakers'—are written in consultation with government officials, and are then transmitted to individuals through the media, and interpreted by national and sub-national legislatures and legal bodies, both IDAs and SYSORs enter the picture as well.

14. For a defence of an objective conception of risk applied to climate change see Shue (2014, Chapter 14).

15. I am not, however, convinced by the way Roberts and Wood try to justify hetero-regulation. The chief problem is how to distinguish the latter from heteronomy. Roberts and Wood argue that for this to happen the directives of the hetero-regulator must be 'assimilated or appropriated' by the agent, and they posit three conditions for such appropriation. First, the agent must understand the terms of the hetero-regulator such that she can 'go on' to 'make creative use' of its directives. This involves knowing both the view expressed by the hetero-regulator and what might motivate opposition to it. Second, the agent must use the hetero-regulator's directives in a habitual or spontaneous manner. And finally, the agent must incorporate the hetero-regulator's norms and directives into her will. That is, she must care about furthering the aims of the hetero-regulator.

The problem has to do with whether or not the ends of the hetero-regulator are themselves capable of endorsement by autonomous epistemic agents. It is crucial to note that this is not itself an issue of reliance on experts because it is a moral issue, not an epistemic one. Imagine an authoritarian and patriarchal religious sect that prescribes beliefs about the inferiority of women. Agents could endorse this sort of hetero-regulator in the ways stipulated by Roberts and Wood: they could make creative use of its directives (including being aware of what might motivate opposition to it), use the latter in a spontaneous manner and care about furthering the aims of the hetero-regulator. Even those who were subject to harsh treatment within the sect could meet these conditions. Some women, though subject to the abuses of the system, might nevertheless endorse it in these ways. But no woman can non-heteronomously endorse her own oppression, from which it follows that fulfilling Roberts' and Wood's three conditions is compatible with exhibiting the *vice* of heteronomy. If the ends of the hetero-regulator do not admit of higher-order justification, then to internalize the beliefs it prescribes is unjustified.

16. The general problem has attracted increasing attention from epistemologists recently. See Coady (1992); Lackey and Sosa (2006); Lackey (2010); Trinkhaus Zagzebski (2012).

17. We might also consider Climategate in this context. Although some have claimed that the hacked emails leading to the scandal at the University of East Anglia's Climatic Research Unit in 2009 revealed collusion among scientists to falsify data, three separate investigations failed to corroborate the charge (which is absurd on its face). See Pearce (2010) for a comprehensive account of these events.

18. The procedures governing the production of IPCC documents can be found in IPCC (1999).

19. Philosophers who prefer this type of explanation often seek to explain self-deception by analogy with interpersonal deception, but this is where the puzzles are thought to arise. For how can one and the same agent be both deceiver and deceived? See Mele (2001, 51). This and the next paragraph are modified from Williston (2002).

20. For a brief discussion of the manner in which the anti-agency view overcomes the puzzles of the agency view see Williston (2002, 69–71).

21. See Spross (2014).

22. Shue (2010, 104). This is the 'liability' view of responsibility for historical emissions. In this context I am agnostic about whether we should adopt a 'strict liability' or 'fault-based liability' version of the claim. According to the former, all countries are responsible for accepting some mitigation costs in proportion to their historical emissions; according to the latter this is only the case above a baseline of 'survival' emissions (below this level emissions are cost-free). The distinction matters because on the strict liability view India, for example, would be required to accept *some* mitigation costs (since it is responsible for some emissions), while on the fault-based view it would not (since on a per-capita basis it remains at or below survival level). A careful treatment of these issues, and a defence of the fault-based view, is in Vanderheiden (2008, 71–8). Caney has argued that if those who are causally responsible for a state of affairs are asked to bear a disproportionate share of the repair costs, they might justifiably see this as a form of punishment. See Caney (2010, 131).

23. For example, Nevada Governor Brian Sandoval (R) recently signed a bill to eliminate 800 megawatts of coal-powered electricity and added 350 megawatts of renewable energy to the State system. Similar measures have been undertaken by Michigan Governor Rick Snyder (R). Both politicians consistently dodge questions about who is to blame for climate change (Spross 2014).

24. These points are meant to cover criteria (1) and (4) of Mele's four conditions. Because I think they are the toughest of the four to explain in this case, I assume the other two are not problematic.

25. My way of talking about the polluter pays principle here is rooted in Scanlon's contractualism. The theory has it that a moral proposition's bindingness has to do with the extent to which a reasonable agent could reject it. If she cannot she is morally required to act in accordance with it. Importantly, the agent in question here is one who takes seriously the impacts of her actions on others. Even if a proposed action will saddle her with costs it may be unreasonable to reject it if doing so imposes even greater costs on others. So there is a respect for the welfare of other agents built into the theory about how best to conduct our moral deliberations. What I am suggesting is that this way of treating moral propositions accords them all the 'objectivity' we need to speak intelligibly of moral truth and falsity. Their objectivity, that is, resides in significant part in the well-being of other reasonable agents, the ones who will be affected by my actions and are thus in effect parties to a contract with me. Of course, there are many subtleties and complexities in Scanlon's contractualism—for example, his critique of aggregative utilitarianism—that go well beyond my purposes here. See Scanlon (2000).

26. This is not to suggest that the PPP cannot be refined. A good analysis of the complexities of PPP is in Caney (2010). Although Caney is critical of PPP, he thinks it needs supplementation (by an 'ability to pay' consideration) rather than replacement. See Vanderheiden (2008) for a similar claim.

27. See Plato (1981, 50a–54e). Socrates argues here that since he has been nourished and protected by his city, the embodiment of which is the law, it would be 'impious' to refuse to submit to its judgment now. This is because if he did refuse, he would be destroying that which has benefited him, even though he has voluntarily accepted the benefits his whole life. On the other hand, if the city decides to destroy Socrates he must either persuade it to do otherwise or submit to its decision. For similar arguments see Shue (2010, 105).

28. What this means, of course, is that I am effectively arguing that self-deception requires Mele's four conditions *as well as* the sort of anxious flight described by Barnes. It would take some space to work this up into a general theory of self-deception, something I cannot do here, but the suggestion strikes me as prima facie plausible.

29. Much of Gardiner's work, especially his analysis of our 'moral corruption', takes this approach. See Gardiner (2012, Part E).

Chapter 6

1. See, for example, Monbiot (2006); Lynas (2008); Hansen (2009); Dyer (2008); Brand (2010).

2. See, for example, Downie (1963); Day (1969, 1970); Bovens (1999); Gravlee (2000); Dodd (2004); Moellendorf (2006); Petit (2004); Steinbock (2007); Meirav (2009); McGeer (2008).

3. There are two exceptions. The first is Nelson (2010). The second is Thompson (2010) who has recently written about the connection between climate change

and radical hope, my subject in part of this chapter. Thompson's article is insightful, but our focuses are significantly different. He is concerned with the narrow problem of how we can 'go on as environmentalists' in the face of the climate crisis. More specifically, he thinks we need radical hope for a future in which we will have to relate to nature as something stripped of all autonomy.

4. See Meirav (2009, 219); also Day (1969) and (1970).
5. See Martin (2014, 48).
6. Nelson (2010) misses this point in his analysis of hope, arguing that the problem with hope is that it is too invested in consequences or outcomes. He claims that because of this hope cannot motivate us adequately in the face of the threat of climate change. We should instead simply act on the basis of moral principles or green virtues, with no view to how things might turn out. This way of seeing things is not only contradicted by the psychological literature, it also seems to confuse hope with wishful thinking. Moreover, it is plausible to claim that we should *both* act on the basis of moral principles or green virtues *and* be hopeful about solving the climate crisis.
7. Martin argues that this problem has mainly to do with distinguishing 'hope against hope' (hope directed at extremely important objects in the face of very poor odds, which I discuss just below) from regular hope, but I think it clearly applies more generally. See Martin (2014, 14–17).
8. A harrowing account of despair can be found in Styron (1992).
9. Scheffler thinks that the doomsday hypothesis illuminates two more 'general features of valuing'. It shows that valuing is not entirely experiential, and is largely non-consequentialist. We won't be around to experience the fact of our non-existence (hence non-experientialism is true); and we would bemoan the outcome even if we calculated that the absence of wars, genocides, and injustice it would entail would, on balance, be a good thing (hence the non-consequentialism) (Scheffler (2012, 32).
10. See Newitz (2014).
11. There may of course be other reasons to care about the future. For example, we might be optimistic that life will persist, even if no humans are there to appreciate it. Or, assuming remnants of humanity remain, we might care about the persistence of consciousness. But at least as compared to the normative connection I am talking about it is more difficult to see how to root these cares in our deepest values.
12. This discussion is analogous to the question of how to distribute emissions permits under a cap. As Jamieson has argued, one possibility is that we do so on the basis of 'productivity'. He argues against this approach because, although it is sensible to adopt it for the allocation of permits, we should not use it to determine the initial distribution of permits. However, in a situation of dire scarcity such as the one I am considering here this approach to the issue is inadequate. If we refuse to base our original distribution of 'permits' on considerations of efficiency or productivity we threaten the ability of the group to survive at all, given the current crisis. In fact, what I am arguing for here flips Jamieson's distinction on its head. In population bottlenecks we must begin with productivity considerations and then, where there are subsequent competing interests, we allocate 'permits' on the basis of a fair lottery. See Jamieson (2010, 272).
13. Think of the practice of resource rotation among Canadian First Nations people. To mention just a few examples, the James Bay Cree employ a four-year rotation

schedule for their beaver hunting grounds, a five- to ten-year schedule for lake fish, and an eighty- to one hundred-year schedule for caribou. Ecologists refer to this as an 'optimal foraging method', and it is a superbly refined model of group-constraint geared to seventh-generation planning. See Berkes, Colding, and Folke (2000, 1255).

14. As Pomeranz points out, it is important not to be too starry-eyed about Japan's achievement. The demographic cushion was provided by Japan's incessant warfare in the sixteenth century, which 'left destruction everywhere'. The wars allowed for the unification of Japan under 'what was probably the world's strongest state system of internal control' (Pomeranz 2013, 31).

15. This discussion touches on a distinction made by Goodin about green taxes. Goodin distinguishes between taxes that function to enforce policies arrived at through free political deliberation (based on sound science, where relevant) and taxes that express the willingness to pay of individual consumers and firms. Because of the problem of externalization, the tax in the latter case is unlikely to be adequate. This is why the policy must be constructed on the basis of non-economic—and, I would add, specifically moral—principles. See Goodin (2010, 240–2).

16. If we insist on retaining the language of necessity in this area, we should at least adopt Shue's understanding of it, according to which our emissions 'are a necessary way of life only until we advance beyond fossil fuel energy technology'. Shue refers to this as an 'avoidable necessity'. Although it is an improvement on the bare reference to necessity, to my mind this way of talking is likely to provide dangerous comfort to the forces of economic and political inertia. See Shue (2014, 115–16).

17. Gardiner does recognize that there are strict constraints on appeals to self-defence, but he does not elaborate how they might operate in the climate change case. See Gardiner (2010, 430–1).

18. Similarly, in their stirring account of a climate change–induced collapse of civilization, Naomi Oreskes and Erik M. Conway express hope that the worst outcomes will be avoided (even though the story they tell is pretty bad). But the sort of warming that might lead to full-scale barbarism is avoided in their tale only by the lucky discovery by a Japanese scientist of a lichenized black fungus whose photosynthetic partner is a highly efficient consumer of atmospheric carbon. Against the wishes of his government, the courageous scientist releases the fungus and 'saves the world'. Something like this may happen, but to place confidence in it doing so is an expression of faith, not rational hope. See Oreskes and Conway (2014, 32).

19. This chapter is a revised version of Williston 2012a. I thank Indiana University Press for permission to use it here. IUP reserves all rights.

Chapter 7

1. On the problem of collective agency see Bratman (2014); Tuomela (2013); List and Pettit (2011); Isaacs (2011).

2. See Hanser (1990); also Harman (2004, 2009); and Shriffin (1999).

3. Woolard astutely highlights the tension in our thinking about this issue (2012, 688–9).

4. See Griswold (2007, 187).

Bibliography

Agarwal, Anil and Sunita Narain. 2003. *Global Warming in an Unequal World: A Case of Environmental Colonialism*. New Delhi: Centre for Science and Environment. <http://cseindia.org/agenda2011/pdf/global_warming%20_agarwal%20and%20narain.pdf>. Accessed 26 October 2014.

Agbor, Julius et al. February, 2012. 'Around the Halls: 2012 Senegal Presidential Election.' Brookings Institute. <http://www.brookings.edu/blogs/up-front/posts/2012/02/10-senegal-halls>. Accessed 17 November 2014.

Allenby, Braden R. and Daniel Sarewitz. 2011. *The Techno-Human Condition*. Cambridge, MA: The MIT Press.

Allison, Ian, Nathan Bindorf, et al. 2009. *Copenhagen Diagnosis: Updating the World on the Latest Climate Science*. <http://www.copenhagendiagnosis.com/default.html>. Accessed 19 March, 2015.

Alvarez, Alex. 2009. *Genocidal Crimes*. London: Routledge.

Annas, Julia. 2011. *Intelligent Virtue*. Oxford: Oxford University Press.

Anscombe, G.E.M. 1958. 'Modern Moral Philosophy.' Philosophy 33 (124): pp. 1–19.

Appiah, Kwame Anthony. 2005. *The Ethics of Identity*. Princeton: Princeton University Press.

Appiah, Kwame Anthony. 2010. *The Honor Code*. New York: Norton and Company.

Aristotle. 1947. *Nichomachean Ethics*. New York: Modern Library.

Armstrong, D.M., E.M. Mauk, et al. 2013. 'Global Warming in an Independent Record of the Past 130 Years.' *Geophysical Research Letters* 40 (1): pp. 189–93.

Atkin, Emily. 18 September, 2014. 'Watch Obama's Top Science Advisor Repeatedly Shut Down Climate Deniers at House Climate Hearings'. *Climate Progress*. <http://thinkprogress.org/climate/2014/09/18/3568720/john-holdren-science-house-climate-hearing/>

Bacigalupi, Paolo. 2009. *The Windup Girl*. San Francisco: Night Shade Books.

Baehr, Jason. 2011. *The Inquiring Mind: On Intellectual Virtues and Virtue Epistemology*. Oxford: Oxford University Press.

Bailey, Joe. 1988. *Pessimism*. New York: Routledge.

Barnes, Annette. 1997. *Seeing Through Self-Deception*. Cambridge: Cambridge University Press.

Barnett, Jon and W. Neil Adger. 2007. 'Climate Change, Human Security and Violent Conflict.' *Political Geography* 26: pp. 639–55.

Beck, Linda J. and Mark E. Pires. 2011. 'West Africa I: Cote D-Ivoire, Nigeria and Senegal.' In *Climate Change and National Security: A Country-Level Analysis*, edited by Daniel Moran, pp. 203–20. Washington: Georgetown University Press.

Beck, Ulrich. 2007. *World at Risk*. London: Polity Press.

Beckerman, Wilfred. 1992. 'Global Warming and International Action: An Economic Perspective.' In *The International Politics of the Environment*, edited by Andrew Burrell and Benedict Kingsbury, pp. 253–89. Oxford: Oxford University Press.

Berkes, Fikret, Johan Colding and Carl Folke. 2000. 'Rediscovery of Traditional Ecological Knowledge as Adaptive Management.' *Ecological Applications* 10 (5): pp. 1251–62.

Berners-Lee, Mike and Duncan Clark. 2014. *The Burning Question*. Vancouver: Greystone Books.

Bovens, Luc. 1999. 'The Value of Hope.' *Philosophy and Phenomenological Research* 59 (3): pp. 667–81.

Brand, Stuart. 2010. *Whole Earth Discipline: An Ecopragmatist Manifesto*. New York: Viking Press.

Bratman, Michael P. 2014. *Shared Agency: A Planning Theory of Acting Together*. Oxford: Oxford University Press.

Broecker, Wallace Smith. 2000. 'Abrupt Climate Change: Causal Constraints Provided by the Paleoclimate Record.' *Earth Science Reviews* 51 (1): pp. 137–54.

Broome, John. 2012. *Climate Matters: Ethics in a Warming World*. New York: Norton.

Brown, Donald A. 2013. *Climate Change Ethics: Navigating the Perfect Moral Storm*. London: Routledge.

Brown, James Robert. 2001. *Who Rules in Science? An Opinionated Guide to the Wars*. Cambridge, MA: Harvard University Press.

Brown, Oli, Anne Hammill and Robert Mcleman. 2007. 'Climate Change as the "New" Security Threat: Implications for Africa.' *International Affairs* 83 (6): pp. 1141–54.

Bruckner, Pascal. 2010. *The Tyranny of Guilt: An Essay on Western Masochism*. Princeton: Princeton University Press.

Brulle, Robert J., Jason Carmichael and Craig Jenkins. 2012. 'Shifting Public Opinion on Climate Change: An Empirical Assessment of Factors Influencing Concern over Climate Change in the U.S, 2002–2010.' *Climatic Change*. DOI: 10.1007/s10584-012-0403-y. Accessed 26 November 2014.

Brysse, Keynyn et al. 2013. 'Climate Change Prediction: Erring on the Side of Least Drama.' *Global Environmental Change* 23 (1): pp. 327–37.

Cafaro, Philip. 2011. 'Alternative Wedges to Create Sustainable Societies.' In *The Ethics of Global Climate Change*, edited by Denis G. Arnold, pp. 192–215. Cambridge: Cambridge University Press.

Caney, Simon. 2010. 'Cosmopolitan Justice, Responsibility and Global Climate Change.' In *Climate Ethics: Essential Readings*, edited by Stephen Gardiner, Simon Caney and Henry Shue, pp. 122–45. Oxford: Oxford University Press.

Caputo, John D. 2013. *Truth: Philosophy in Transit*. New York: Penguin Books.

Card, Claudia. 2005. *The Atrocity Paradigm: A Theory of Evil*. Oxford: Oxford University Press.

Cassirer, Ernst. 1951. *The Philosophy of the Enlightenment*. Princeton: Princeton University Press.

Clarkson, Linda et al. 1992. *Our Responsibility to the Seventh Generation: Indigenous Peoples and Sustainable Development*. Winnipeg: International Institute for Sustainable Development.

Coady, C.A.J. 1992. *Testimony: A Philosophical Study*. Oxford: Oxford University Press.

Commodities Now. 13 November 2014. 'U.S. and China Targets are Game Changer for Climate Negotiations.' <http://www.commodities-now.com/news/environmental-markets/17773-us-and-china-targets-are-game-changer-for-climate-negotiations.html?utm_source=Weekly+Carbon+Briefing&utm_campaign=1587b9eb6c-Carbon_Brief_Weekly_111114&utm_medium=email&utm_term=0_3ff5ea836a-1587b9eb6c-303417537>. Accessed 14 November 2014.

Costelloe, Michael T. et al. 2007. 'Punitive Attitudes Towards Criminals: Exploring the Relevance of Crime Salience and Economic Insecurity.' *Punishment and Society* 11: pp. 25–49.

Cripps, Elisabeth. 2013. *Climate Change and the Moral Agent: Individual Duties in an Interdependent World*. Oxford: Oxford University Press.

Cunningham, Anthony. 2013. *Modern Honor: A Philosophical Defence*. London: Routledge.

Daily, Matt. 27 June 2012. 'Exxon CEO Calls Climate Change Engineering Problem.' *Reuters*. <http://www.reuters.com/article/2012/06/27/us-exxon-climate-idUSBRE85Q1C820120627>. Accessed 4 September 2014.

Dalby, Simon. 2013. 'New Dimensions of Climate Security.' *The RUSI Journal* 158 (3): pp. 34–43.

Dalby, Simon. 2009. *Security and Environmental Change*. Cambridge: Polity Press.

Dark Mountain Project. 2014. <http://dark-mountain.net/about/the-dark-mountain-project/>. Accessed 14 June 2014.

Day, J. P. 1970. 'The Anatomy of Hope and Fear.' *Mind* 79 (315): pp. 369–84.

Day, J. P. 1969. 'Hope.' *American Philosophical Quarterly* 6: pp. 89–102.

de Gouges, Olympe. 2013. *Ainsi soit Olympe de Gouges: La déclaration des droits de la femme*. Paris: Grasset et Fasquelle.

de Montaigne, Michel. 1991. *The Complete Essays*. Harmondsworth, UK: Penguin.

de Tocqueville, Alexis. 2010. *The Old Regime and the French Revolution*. New York: Dover.

Déry, Stephen J. and Ross D. Brown. 2007. 'Recent Northern Hemisphere Snow Cover Extent Trends and Implications for the Snow-Albedo Feedback.' *Geophysical Research Letters* 34 (22). DOI: 1029/2007GL031474. Accessed 13 May 2014.

Dodd, James. 2004. 'The Philosophical Significance of Hope.' *The Review of Metaphysics* 58: pp. 117–46.

Doris, John. 2005. *Lack of Character: Personality and Moral Behaviour*. Cambridge: Cambridge University Press.

Downie, R.S. 1963. 'Hope.' *Philosophy and Phenomenological Research* 24 (2): pp. 248–51.

Dumanoski, Dianne. 2011. *The End of the Long Summer: Why We Must Remake our Civilization to Survive on a Volatile Earth*. New York: Crown Publishers.

Dyer, Gwynne. 2008. *Climate Wars*. Toronto: Vintage Canada.

Feldman, Stacey. 26 February 2010. 'Climate scientists defend IPCC peer review as most rigorous in history.' *Inside Climate News*. <http://insideclimatenews.org/news/20100226/climate-scientists-defend-ipcc-peer-review-most-rigorous-history>. Accessed 14 January 2012.

Flannery, Tim. 2011. *Here on Earth: A Natural History of the Planet*. New York: Harper Collins.

Foot, Philippa. *Natural Goodness*. 2001. Oxford: Oxford University Press.

Frankfurt, Harry. 1988. 'On Bullshit.' In *The Importance of What We Care About*. Cambridge: Cambridge University Press, pp. 117–33.

Friel, Howard. 2011. *The Lomborg Deception: Setting the Record Straight about Global Warming*. New Haven: Yale University Press.

Gardiner, Stephen J. 2012. 'Are We the Scum of the Earth? Climate Change, Geoengineering and Humanity's Challenge.' In *Ethical Adaptation to Climate Change: Human Virtues of the Future*, edited by Allen Thompson and Jeremy Bendik-Keymer, pp. 241–60. Cambridge, MA: The MIT Press.

Gardiner, Stephen J. 2010. 'Is "Arming the Future" with Geoengineering Really the Lesser Evil: Some Doubts about Intentionally Manipulating the Climate System.' In *Climate Ethics: Essential Readings*, edited by Stephen J. Gardiner, Simon Caney and Henry Shue, pp. 284–312. Oxford: Oxford University Press.

Gardiner, Stephen J. 2011. *A Perfect Moral Storm: The Ethical Tragedy of Climate Change*. Oxford: Oxford University Press.

George, Rick. 2012. *Sunrise: Suncor, the Oil Sands and the Future of Energy*. Toronto: Harper-Collins.

Giddens, Anthony. 1990. *The Consequences of Modernity*. Stanford: Stanford University Press.

Giddens, Anthony. 1991. *Modernity and Self-Identity: Self and Society in the Late Modern Age*. Stanford: Stanford University Press.

Glover, Jonathan. 2000. *Humanity: A Moral History of the Twentieth-Century*. New Haven: Yale University Press.

Godlovitch, Stan. 1998. 'Morally we Roll Along: (Optimistic Reflections) on Moral Progress.' *Journal of Applied Philosophy* 15 (3): pp. 271–86.

Goldberg, Sanford C. 2011. 'If That Were True I Would Have Heard about it by Now.' In *Social Epistemology: Essential Readings*, edited by Sanford C. Goldman and Denis Whitcomb, pp. 92–108. Oxford: Oxford University Press.

Goldman, Alvin I. 2011. 'A Guide to Social Epistemology.' In *Social Epistemology: Essential Readings*, edited by Sanford C. Goldman and Denis Whitcomb, pp. 11–37. Oxford: Oxford University Press.

Goldstone, Jack A. 2013. 'Climate Lessons from History.' *Historically Speaking* 14 (5): pp. 35–7.

Goodin, Robert E. 2010. 'Selling Environmental Indulgences.' In *Climate Ethics: Essential Readings*, edited by Stephen J. Gardiner, Simon Caney and Henry Shue, pp. 231–46. Oxford: Oxford University Press.

Gourevitch, Philip. 1998. *We Wish to Inform You That Tomorrow We Will Be Killed with Our Families: Stories from Rwanda*. New York: Picador.

Gravlee, Scott. 2000. 'Aristotle on Hope.' *Journal of the History of Philosophy* 38 (4): pp. 461–77.

Greco, John. 2011. 'Virtue Epistemology.' *Stanford Encyclopedia of Philosophy*. <http://plato.stanford.edu/entries/epistemology-virtue/>. Accessed 19 March 2015.

Gribbin, John and Mary Gribbin. 2009. *He Knew He Was Right: The Irrepressible Life of James Lovelock*. London: Penguin.

Griswold, Charles. 1998. *Adam Smith and the Virtues of Enlightenment*. Cambridge: Cambridge University Press.

Griswold, Charles. 2007. *Forgiveness: A Philosophical Exploration*. Cambridge: Cambridge University Press.

Hamilton, Clive. 2014. 'The Delusion of the "Good Anthropocene": Reply to Andrew Revkin.' <http://clivehamilton.com/the-delusion-of-the-good-anthropocene-reply-to-andrew-revkin/>. Accessed 21 November 2014.

Hamilton, Clive. 2013. 'What History Can Teach Us about Climate Change Denial.' In *Engaging with Climate Change: Psychoanalytic and Interdisciplinary Perspectives*, edited by Sally Weintrobe, pp. 16–32. London: Routledge.

Hamilton, Clive. 2010. *Requiem for a Species: Why We Resist the Truth About Climate Change*. London: Taylor and Francis.

Hansen, James. 2009. *Storms of My Grandchildren: The Truth about the Coming Climate Catastrophe and our Last Chance to Save Humanity*. New York: Bloomsbury.

Hansen, James et al. 2008. 'Target Atmospheric CO2: Where should Humanity Aim?' *The Open Atmospheric Science Journal* 2 (1): pp. 217–31. DOI: 10.2174/1874282300802010217. Accessed 12 March 2009.

Hanser, Matthew. 1990. 'Harming Future People.' *Philosophy and Public Affairs* 19: pp. 47–70.

Harman, E. 2009. 'Harming as Causing Harm.' In *Harming Future People: Ethics, Genetics and the Non-Identity Problem*, edited by M.A. Roberts and D.T. Wasserman, pp. 137–54. Dordrecht: Springer.

Harman, Elizabeth. 2004. 'Can We Harm and Benefit in Creating?' *Philosophical Perspectives* 18: pp. 89–113.

Harman, Gilbert. 1999. 'Moral Philosophy Meets Social Psychology: Virtue Ethics and the Fundamental Attribution Error.' In *Proceedings of the Aristotelian Society* 99: pp. 315–31.

Harman, Gilbert. 2000. 'The Nonexistence of Character Traits.' In *Proceedings of the Aristotelian Society* 100: pp. 223–6.

Harris, George. 2006. *Reason's Grief: An Essay on Tragedy and Value*. Cambridge: Cambridge University Press.

Heath, Joseph. 2014. *Enlightenment 2.0: Restoring Sanity to our Politics, Our Economy and our Lives*. Toronto: Harper Collins.

Hedges, Chris. 2010. *Death of the Liberal Class*. Toronto: Vintage Canada.

Heilbroner, Robert. 1981. *An Inquiry into the Human Prospect*. New York: Norton and Company.

Herzog, T., J. Pershing and K.A. Baumert. 2005. *Navigating the Numbers: Greenhouse Gas Data and International Climate Policy*. World Resources Institute. <http://www.wri.org/publication/navigating-numbers>. Accessed 4 May 2011.

Hill, Thomas E. Jr. 1983. 'Ideals of Human Excellence and Preserving Natural Environments.' *Environmental Ethics* 5 (3), 211–224.

Hirsch, Paul D. and Brian G. Norton. 2012. 'Thinking Like a Planet.' In *Ethical Adaptation to Climate Change: Human Virtues of the Future*, edited by Allen Thompson and Jeremy Bendik-Keymer, pp. 317–34. Cambridge, MA: The MIT Press.

Hirschman, Albert O. 1977. *The Passions and the Interests: Political Arguments for Capitalism before its Triumph*. Princeton: Princeton University Press.

Hiskes, Richard P. 2008. *The Human Right to a Green Future: Environmental Rights and Intergenerational Justice*. Cambridge: Cambridge University Press.

Hobbes, Thomas. 1994. *Leviathan*, edited by Edwin Curley. Indianapolis: Hackett.

Hoffmann, Tobias. 2008. *Weakness of Will from Plato to the Present*. Washington: Catholic University of America Press.

Hoffmeister, John. 2010. *Why We Hate the Oil Companies*. New York: Palgrave MacMillan.

Hoggan, James. 2009. *Climate Cover-Up: The Crusade to Deny Global Warming*. Vancouver: Greystone Books.

Holmgren, Margaret R. 2014. *Forgiveness and Retribution*. Cambridge: Cambridge University Press.

Homer-Dixon, Thomas. 1994. 'Environmental Scarcities and Conflict: Evidence from Cases.' *International Security* 19 (1): pp. 5–40.

Hulme, Mike. 2009. *Why We Disagree about Climate Change: Understanding Controversy, Inaction and Opportunity*. Cambridge: Cambridge University Press.

Hume, David. 1882. 'Of National Characters.' In *The Political Works of David Hume*, edited by T.H. Green and T.H. Grose. London: Longrass, Green and Company.

Hursthouse, Rosalind. 1999. *On Virtue Ethics*. Oxford: Oxford University Press.

Hursthouse, Rosalind. 1991. 'Virtue Ethics and Abortion.' *Philosophy and Public Affairs* 20 (3): pp. 223–46.

Inhofe, James. 2011. 'The Science of Climate Change: Senate Floor Statement.' In *The Global Warming Reader*, edited by Bill McKibben, pp. 164–91. New York: Penguin Books.

Innerarity, Daniel. 2012. *The Future and Its Enemies: In Defense of Political Hope*. Stanford: Stanford University Press.

Intergovernmental Panel on Climate Change (IPCC). 2014a. *Working Group II: Impacts, Adaptation and Vulnerability*. <http://www.ipcc.ch/pdf/assessment-report/ar5/wg2/WGIIAR5-Chap12_FINAL.pdf>. Accessed 3 November 2014.

International Energy Agency (IEA). 2013. *World Energy Outlook 2013 Factsheet*. <http://www.worldenergyoutlook.org/media/weowebsite/factsheets/WEO2013_Factsheets.pdf>. Accessed 9 November 2013.

International Geosphere-Biosphere Program (IGBP). 2001. *Amsterdam Declaration on Earth System Science*. <http://www.igbp.net/about/history/2001amsterdamdeclarationonearthsystemscience.4.1b8ae20512db692f2a680001312.html>. Accessed 13 October 2014.

IPCC. 2007. 'Climate Change and its impacts in the near and long term under different scenarios.' *Synthesis Report*, Chapter 3. <http://www.ipcc.ch/publications_and_data/ar4/syr/en/spms3.html>. Accessed 19 April 2009.

IPCC. 1999. 'Procedures for the Preparation, Review, Acceptance, Adoption, Approval and Adoption of IPCC Reports.' *Appendix A to the Principles Governing IPCC Work*. <https://www.ipcc.ch/pdf/ipcc-principles/ipcc-principles-appendix-a.pdf>. Accessed 14 July 2010.

IPCC. 2014b. *Working Group III: Mitigation of Climate Change*. <http://mitigation2014.org/report>. Accessed 2 November 2014.

Isaacs, Tracy. 2011. *Moral Responsibility in Collective Contexts*. Oxford: Oxford University Press.

Jackson, Tim. 2011. *Prosperity Without Growth: Economics for a Finite Planet*. London: Earthscan.

Jameson, Frederic. 1992. *Postmodernism, Or the Cultural Logic of Late Capitalism*. Durham, NC: Duke University Press.

Jamieson, Dale. 2010. 'Adaptation, Mitigation and Justice.' In *Climate Ethics: Essential Readings*, edited by Stephen J. Gardiner, Simon Caney and Henry Shue, pp. 263–83. Oxford: Oxford University Press.

Jamieson, Dale. 2012. 'Ethics, Public Policy and Global Warming.' In *Ethical Adaptation to Climate Change: Human Virtues of the Future*, edited by Allen Thompson and Jeremy Bendik-Keymer, pp. 187–202. Cambridge, MA: The MIT Press.

Jamieson, Dale. 2002. *Morality's Progress: Essays on Humans, Other Animals and the Rest of Nature*. Oxford: Clarendon Press.

Jamieson, Dale. 2014. *Reason in a Dark Time: Why the Struggle against Climate Change Failed and What it Means for Our Future*. Oxford: Oxford University Press.

Jamieson, Dale. 2007. 'When Consequentialists Should Be Virtue Theorists.' *Utilitas* 19 (2): pp. 1–24.

Kant, Immanuel. 2011. 'The Difference between the Races.' In *Observations on the Feeling of the Beautiful and the Sublime*, pp.11–205. Cambridge: Cambridge University Press.

Kant, Immanuel. 1983. 'Preface and Introduction to the Metaphysics of Morals.' In *Immanuel Kant: Ethical Philosophy*, translated by James Ellington. Indianapolis: Hackett Publishing, Book II, 1–30.

Kant, Immanuel. 1988. 'What is Enlightenment?' In *Kant Selections*, edited by Lewis White Beck. New York: Scribner/Macmillan Publishing, 462–7.

Kawall, Jason. 2012. 'Rethinking Greed.' In *Ethical Adaptation to Climate Change: Human Virtues of the Future*, edited by Allen Thompson and Jeremy Bendik-Keymer, pp. 223–40. Cambridge, MA: The MIT Press.

Klare, Michael T. 2012. *The Race for What's Left: The Global Scramble for the World's Last Resources*. New York: Henry Holt and Company.

Kornblith, Hilary. 2001. *Epistemology: Internalism and Externalism*. London: Blackwell.

Korsgaard, Christine M. 2009. *Self-Constitution: Agency, Identity and Integrity*. Oxford: Oxford University Press.

Korsgaard, Christine M. 1996. *The Sources of Normativity*. Cambridge: Cambridge University Press.

Lackey, Jennifer. 2010. *Learning from Words: Testimony as a Source of Knowledge*. Oxford: Oxford University Press.

Lackey, Jennifer and Ernst Sosa (editors). 2006. *The Epistemology of Testimony*. Oxford: Oxford University Press.

Lane, Melissa. 2012. *Eco-Republic: What the Ancients can Teach us about Ethics, Virtue and Sustainable Living*. Princeton: Princeton University Press.

Lasch, Christopher. 1985. *The Minimal Self: Psychic Survival in Troubled Times*. New York: WW Norton.

Lear. Jonathan. 2006. *Radical Hope: Ethics in the Face of Cultural Devastation*. Cambridge, MA: Harvard University Press.

Lee, James R. 2011. *Climate Change and Armed Conflict: Hot and Cold Wars*. London: Routledge.

Lewis, David. 2000. 'Dispositional Theories of Value.' In *Papers in Ethics and Social Philosophy*, pp. 68–94. Cambridge: Cambridge University Press.

Lewis, Joanna I. 2011. 'China.' In *Climate Change and National Security: A Country-Level Analysis*, edited by Daniel Moran, pp. 9–26. Washington: Georgetown University Press.

List, Christian and Philip Pettit. 2011. *The Possibility, Design and Status of Corporate Agents*. Oxford: Oxford University Press.

Lloyd, Genevieve. 2013. *Enlightenment Shadows*. Oxford: Oxford University Press.

Lomborg, Bjørn. 2010. *Cool It: The Skeptical Environmentalist's Guide to Global Warming*. New York: Vintage.

Lomborg, Bjørn. 2001. The *Skeptical Environmentalist*. Cambridge: Cambridge University Press.

Lovelock, James. 2006. *The Revenge of Gaia*. London: Penguin Books.

Lovelock, James. 2014. *A Rough Ride to the Future*. London: Allen Lane.

Lynas, Mark. 2008. *Six Degrees: Our Future on a Hotter Planet*. London: Harper Perennial.

MacDonald, Gayle. 1 May 2014. 'Youth Anxiety on the Rise amid Climate Change.' *Globe and Mail*. <http://www.theglobeandmail.com/life/health-and-fitness/health/youth-anxiety-on-the-rise-amid-changing-climate/article18372258/>. Accessed 3 August 2014.

Magdoff, Fred and John Bellamy Foster. 2011. *What Every Environmentalist Needs to Know about Capitalism*. New York: Monthly Review Press.

Mann, Michael E. 2009. 'Defining Dangerous Anthropogenic Interference.' *Proceedings of the National Academy of Sciences* 106 (4): pp. 4065–6.

Martin, Adrienne M. 2014. *How We Hope: A Moral Psychology*. Princeton: Princeton University Press.

Marx, Karl. 1964. *Economic and Philosophic Manuscripts of 1844*. New York: International Publishers.

McEwan, Ian. 2010. *Solar*. Toronto: Vintage Canada.

McGeer, Victoria. 2008. 'Trust, Hope and Empowerment.' *Australasian Journal of Philosophy* 86 (2): pp. 237–54.

McGrath, Matt. 11 November 2013. 'Typhoon prompts "fast" by Philippines climate delegate.' *BBC News*. <http://www.bbc.co.uk/news/science-environment-24899647>. Accessed 20 November 2013.

McKibben, Bill. 19 July 2012. 'Global Warming's Terrifying New Math.' *Rolling Stone Magazine*. <http://www.rollingstone.com/politics/news/global-warmings-terrifying-new-math-20120719>. Accessed 28 July 2012.

McKibben, Bill. 2012. *The Global Warming Reader: A Century of Writing About Climate Change*. London: Penguin Books.

McKinnon, Catriona. 2011. *Climate Change and Future Justice: Precaution, Compensation and Triage*. London: Routledge.

Meadows, Donella. H. 2009. *Thinking in Systems: A Primer*. London: Routledge.

Meirav, Ariel. 2009. 'The Nature of Hope.' *Ratio XXII* (2): pp. 216–33.

Mele, Alfred. 2001. *Self-Deception Unmasked*. Princeton: Princeton University Press.

Miller, David. 2010. 'Cosmopolitanism.' In *The Cosmopolitanism Reader*, edited by Garrett Wallace Brown and David Held, pp. 377–92. Cambridge: Polity Press.

Miller, William Ian. 2000. *The Mystery of Courage*. Cambridge, MA: Harvard University Press.

Milly, P.C.D. et al. 2008. 'Stationarity is Dead: Wither Water Management?' *Science* 319: pp. 573–4.

Moellendorf, Dale. 2006. 'Hope as a Political Virtue.' *Philosophical Papers* 35 (3): pp. 413–33.

Moellendorf, Dale. 2014. *The Moral Challenge of Dangerous Climate Change*. Cambridge: Cambridge University Press.

Monbiot, George. 2006. *Heat: How to Stop the Planet from Burning*. Toronto: Doubleday Canada.

Montmarquet, James. 1987. 'Epistemic Virtue.' *Mind* 96: pp. 482–97.

Moody-Adams, Michelle M. 1999. 'The Idea of Moral Progress.' *Metaphilosophy* 30 (3): pp. 168–85.

Mora, C. et al. 2013. 'The Projected Timing of Climate Departure from Recent Variability.' *Nature* 502: pp. 183–7.

Moran, Daniel (ed.). 2011. *Climate Change and National Security: A Country-Level Analysis*. Washington: Georgetown University Press.

Morgan, Michael L. 2008. *On Shame*. London: Routledge.

Mulgan, Tim. 2011. *Ethics for a Broken World: Imagining Philosophy after Catastrophe*. Montreal: McGill-Queens University Press.

National Academy of Sciences. 2009. 'Restructuring Federal Climate Research to Meet the Challenges of Climate Change.' <http://www.nap.edu/catalog/12595/restructuring-federal-climate-research-to-meet-the-challenges-of-climate-change>. Accessed 12 June 2014.

Neiman, Susan. 2002. *Evil in Modern Thought: An Alternative History of Philosophy*. Princeton: Princeton University Press.

Nelson, Michael P. 2010. 'To A Future Without Hope.' In *Moral Ground: Ethical Action for a Planet in Peril*, edited by Kathleen Moore and Michael P. Nelson, pp. 458–62. San Antonio: Trinity University Press.

Newitz, Annalee. 2014. *Scatter, Adapt and Remember: How Humans Will Survive a Mass Extinction*. New York: Doubleday.

Nietzsche, Friedrich. 1992. *The Birth of Tragedy*. In *Basic Writings of Nietzsche*, translated by Walter Kaufman, pp. 1–144. New York: Random House.

Norgaard, Kari Marie. 2011. *Living in Denial: Climate Change, Emotions, and Everyday Life*. Cambridge, MA: The MIT Press.

Nussbaum, Martha. 2010. 'Patriotism and Cosmopolitanism.' In Brown, Garrett Wallace and David Held. *The Cosmopolitan Reader*. London: Polity Press, 155–62.

O'Neill, Onora. 1996. *Towards Justice and Virtue: A Constructivist Account of Practical Reasoning*. Cambridge: Cambridge University Press.

O'Neill, Saffron and Sophie Nicholson Cole. 2009. 'Fear Won't Do It.' *Science Communication* 30 (3): pp. 355–79.

Oreskes, Naomi and Eric M. Conway. 2014. *The Collapse of Western Civilization: A View from the Future*. New York: Columbia University Press.

Oreskes, Naomi and Eric M. Conway. 2010. *Merchants of Doubt*. London: Bloomsbury Press.

Owen, David. 2011. *The Conundrum: How Scientific Innovation, Increased Efficiency, and Good Intentions Can Make Our Energy and Climate Problems Worse*. New York: Riverhead Books.

Parfit, Derek. 2010. 'Energy Policy and the Further Future: The Identity Problem.' In *Climate Ethics: Essential Readings*, edited by Stephen J. Gardiner, Simon Caney and Henry Shue, pp. 112–21. Oxford: Oxford University Press.

Parfit, Derek. 1984. *Reasons and Persons*. Oxford: Oxford University Press.

Parker, Geoffrey. 2013. *Global Crisis: War, Climate Change and Catastrophe in the Seventeenth-Century*. New Haven: Yale University Press.

Parr, Adrian. 2012. *The Wrath of Capital: Neoliberalism and Climate Politics*. Cambridge: Cambridge University Press.

Paul, T.V. 2011. 'India.' In *Climate Change and National Security: A Country-Level Analysis*, edited by Daniel Moran, pp. 73–84. Washington: Georgetown University Press.

Pearce, Fred. 2010. *The Climate Files: The Battle for the Truth about Global Warming*. London: Guardian Books.

Pears, David. 1984. *Motivated Irrationality*. Oxford: Oxford University Press.

Petit, Philip. 2004. 'Hope and Its Place in Mind.' *Annals of the American Academy of Political and Social Science* 592: pp. 152–65.

Plato. 1981. *Crito*. In *Five Dialogues*, translated by G.M.A. Grube. Indianapolis: Hackett, pp. 45–56.

Plato. 1992. *Republic*, translated by G.M.A. Grube. Indianapolis: Hackett.

Pogge, Thomas. 1992. 'Cosmopolitanism and Sovereignty.' *Ethics* 103: pp. 48–75.

Pomeranz, Kenneth. 2013. 'Weather, War, and Welfare: Persistence and Change in Geoffrey Parker's *Global Crisis*.' *Historically Speaking* 14 (5): pp. 30–3.

Posner, Eric A. and David Weisbach. 2010. *Climate Change Justice*. Princeton: Princeton University Press.

Preston, Christopher J. (editor). 2013. *Engineering the Climate: The Ethics of Solar Radiation Management*. Lanham: Lexington Books.

PricewaterhouseCoopers. 2014. *Two Degrees of Separation: Ambition and Reality*. http://www.pwc.co.uk/assets/pdf/low-carbon-economy-index-2014.pdf.

Public Interest Research Centre (UK). 2013. *Common Cause for Nature: Finding Values and Frames in the Conservation Sector*. <http://valuesandframes.org/>. Accessed 1 December 2013.

Rahmstorf, Stefan et al. 2012. 'Comparing Climate Projection to Observations up to 2011.' *Environmental Research Letters* 7 (4). DOI: 10.1088/1748-9326/7/4/044035. Accessed 18 January 2014.

Randall, R. 2013. 'Great Expectations: The Psychodynamics of Ecological Debt.' In *Engaging with Climate Change: Psychoanalytic and Interdisciplinary Perspectives*, edited by Sally Weintrobe, pp. 87–108. London: Routledge.

Riahi, Keywan, Elmar Kriegler, et al. 2013. 'Locked into Copenhagen Pledges—Implications of short-term emission targets for the cost and feasibility of long-term climate goals.' *Technological Forecasting & Social Change*. Volume 90, Part A, 8-23. <https://www.pik-potsdam.de/research/climate-impacts-and-vulnerabilities/projects/project-pages/world-bank-report/publications/riahi-ampere-delayed-action-tfsc14.pdf>. Accessed 12 April 2014.

Riaz, Ali. 2011. 'Bangladesh.' In *Climate Change and National Security: A Country-Level Analysis*, edited by Daniel Moran, pp. 103–14. Washington: Georgetown University Press.

Robbins, Bruce W. 2012. *Perpetual War: Cosmopolitanism from the Viewpoint of Violence*. Durham: Duke University Press.

Roberts, Robert and W. Jay Wood. 2007. *Intellectual Virtue: An Essay in Regulative Epistemology*. Oxford: Clarendon Press.

Rogers, Simon and Lisa Evans. 2011. 'World carbon dioxide emissions by country: China Speeds Ahead of the Rest.' 31 January 2011, *The Guardian*. <http://www.theguardian. com/news/datablog/2011/jan/31/world-carbon-dioxide-emissions-country-data-co2#data>. Accessed 14 January 2012.

Romm, Joe. 11 August 2011. 'Arctic Ice Thinning Four Times Faster than Predicted by IPCC Models, Semi-Stunning MIT Study Finds.' <http://thinkprogress.org/romm/ 2011/08/11/294403/arctic-ice-thinning-4-times-faster-than-predicted-by-models-semi-stunning-m-i-t-study-finds/>. Accessed 20 October 2011.

Romm, Joe. 18 August 2013. 'New IPCC Report: Climatologists More Certain Global Warming is Caused by Humans, Impacts are Speeding Up.' <http://thinkprogress. org/climate/2013/08/18/2484711/ipcc-report-more-certain-global-warming-is-caused-by-humans-impacts-speeding-up/>. Accessed 11 September 2013.

Romm, Joe. 31 January 2013. 'Why Climate Scientists Have Consistently Underesti-mated Key Global Warming Impacts.' <http://thinkprogress.org/climate/2013/01/31/ 1524981/why-climate-scientists-have-consistently-underestimated-key-global-warm ing-impacts/>. Accessed 12 March 2013.

Rorty, Richard. 2006. 'Philosophy's Role Vis-à-Vis Business Ethics.' *Business Ethics Quarterly* 16 (3): pp. 369–80.

Rubin, Jeff. 2009. 'Demand Shift.' In *Carbon Shift: How the Twin Crises of Oil Depletion and Climate Change will Define the Future*, edited by Thomas Homer-Dixon, pp. 136–52. Toronto: Random House Canada.

Ruddiman. William F. 2005. *Plows, Plagues, and Petroleum: How Humans Took Control of Climate*. Princeton: Princeton University Press.

Rustin, M. 2013. 'Discussion: Great Expectations: The Psychodynamics of Ecological Debt.' In *Engaging with Climate Change: Psychoanalytic and Interdisciplinary Perspec-tives*, edited by Sally Weintrobe, pp. 103–8. London: Routledge.

Sandel, Michael. 2009. *Justice: What's the Right Thing to Do?* New York: Farrar, Straus and Giroux.

Sandler, Ronald. 2007. *Character and Environment: A Virtue-Oriented Approach to Envir-onmental Ethics*. New York: Columbia University Press.

Sandler, Ronald. 2010. 'Ethical Theory and the Problem of Inconsequentialism: Why Environmental Ethicists should be Virtue-Oriented Ethicists.' *Journal of Agricultural and Environmental Ethics* 23: pp. 167–83.

Saramago, José. 1995. *Blindness*. New York: Houghton Mifflin Harcourt.

Scanlon, T.M. 2000. *What We Owe to Each Other*. Cambridge, MA: Harvard University Press.

Scheffler, Samuel. 2012. *Death and the Afterlife*. Oxford: Oxford University Press.

Schellnhuber, H.J. 2009. 'Terra Quasi-Incognito: Beyond the 2°C Line.' *International Climate Conference*. <http://www.eci.ox.ac.uk/4degrees/ppt/1-1schellnhuber.pdf>. Accessed 3 March 2010.

Schönfeld, Martin. 2000. *The Philosophy of the Young Kant: The Pre-Critical Project*. Oxford: Oxford University Press.

Schönfeld, Martin. 1992. 'Who or What Has Moral Standing?' *American Philosophical Quarterly* 29 (4): pp. 353–62.

Semeniuk, Ivan. 31 March 2014. 'New climate change report details threats to global security, possibilities of violent conflict.' *The Globe and Mail*. <http://www.

theglobeandmail.com/news/national/new-climate-change-report-details-threats-to-global-security-possibilities-of-violent-conflict/article17734823/>. Accessed 4 April 2014.

Shakespeare, William. 1995. *A Midsummer's Night Dream*, edited by Peter Holland. Oxford: Oxford University Press.

Shell, S. M. 2001. 'Émile: Nature and the Education of Sophie.' *Cambridge Companion to Rousseau*, edited by Patrick Riley. Cambridge: Cambridge University Press, Chapter 10.

Sherkoske, Gregory. 2012. *Integrity and the Virtues of Reason: Leading a Convincing Life*. Cambridge: Cambridge University Press.

Shklar, Judith. N. 1990. *The Faces of Injustice*. New Haven: Yale University Press.

Shriffin, Seana. 1999. 'Wrongful Life, Procreative Responsibility, and the Significance of Harm.' *Legal Theory* 5: pp. 117–48.

Shue, Henry. 2013. 'Climate Hope: Implementing the Exit Strategy.' *Chicago Journal of International Law* 13 (2): 379–400.

Shue, Henry. 2014. *Climate Justice: Vulnerability and Protection*. Oxford: Oxford University Press.

Shue, Henry. 2010. 'Global Environment and International Inequality.' In *Climate Ethics: Essential Readings*, edited by Stephen J. Gardiner, Simon Caney and Henry Shue, pp. 101–11. Oxford: Oxford University Press.

Simon, Herbert A. 1955. 'A Behavioral Model of Rational Choice.' *Quarterly Journal of Economics* 69: pp. 99–118.

Simon, Julian. L. 1998. *The Ultimate Resource*. Princeton: Princeton University Press.

Singer, Peter. 1981. *The Expanding Circle: Ethics and Sociobiology*. New York: The New American Library.

Slote, Michael. 2001. *Morals from Motives*. Oxford: Oxford University Press.

Slote, Michael. 1997. 'Virtue Ethics.' In *Three Methods of Ethics*, edited by Marcia W. Baron, Philip Pettit and Michael Slote, pp. 175–239. London: Blackwell Publishing.

Slovic, Paul. 2007. 'If I look at the Mass I Will Never Act: Psychic Numbing and Genocide.' *Judgment and Decision Making* 2: pp. 79–95.

Slovic, Paul. 2013. 'Psychic Numbing and Mass Atrocity.' In *The Behavioural Foundations of Public Policy*, edited by E. Shafir, pp. 126–42. Princeton: Princeton University Press.

Snow, Nancy. E. 2010. *Virtue as Social Intelligence: An Empirically Grounded Theory*. New York: Routledge.

Snyder, C. R. and K.L. Rand. 2003. 'The Case Against False Hope.' *American Psychologist* 58: 820–2.

Solokov, A.P., P.H. Stone et al. 2009. 'Probabilistic Forecast for 21st Century Climate Based on Uncertainties in Emissions (without Policy) and Climate Parameters.' *MIT Joint Program on the Science and Policy of Climate Change*. <http://globalchange.mit.edu/research/publications/990>. Accessed 1 December 2014.

Solomon, Robert C. 2007. *True to Our Feelings: What our Emotions are Really Telling Us*. Oxford: Oxford University Press.

Sparshott, Francis. 1994. *Taking Life Seriously: A Study of the Argument of the Nichomachean Ethics*. Toronto: University of Toronto Press.

Spross, Jeff. 2 October 2014. 'Trying to Fight Climate Change without Admitting we are to Blame.' *Think Progress.* <http://thinkprogress.org/climate/2014/10/02/3571238/fighting-climate-change-without-blame/?elq=~~eloqua..type–emailfield..syntax–recipientid~~&elqCampaignId=~~eloqua..type–campaign..campaignid–0..fieldname–id~~>. Accessed October 3, 2014.

Sreenivasan, Gopal. 2002. 'Errors about Errors: Virtue Theory and Trait Attribution.' *Mind* 111: pp. 47–68.

Steffen, Will et al. 2004. 'Abrupt Changes: The Achilles Heels of the Earth System.' *Environment* 46 (3): pp. 9–20.

Steinbock, Anthony J. 2007. 'The Phenomenology of Despair.' *International Journal of Philosophical Studies* 15 (3): pp. 435–51.

Strawson, Peter F. 2008. *Freedom and Resentment and Other Essays.* London: Routledge.

Styron, William. 1992. *Darkness Visible: A Memoir of Madness.* New York: Vintage Press.

Tan, Kok-Chor. 2004. *Justice Without Borders: Cosmopolitanism, Nationalism and Patriotism.* Cambridge: Cambridge University Press.

Taylor, Gabrielle. 2008. *Deadly Vices.* Oxford: Oxford University Press.

Taylor, Gabrielle. 1985. *Pride, Shame and Guilt.* Oxford: Oxford University Press.

Tessman, Lisa. 2005. *Burdened Virtues: Virtue Ethics for Liberatory Struggles.* Oxford: Oxford University Press.

Thompson, Allen. 2010. 'Radical Hope for Living Well in a Warming World.' *Journal of Agricultural and Environmental Ethics* 23 (1): pp. 43–59.

Trinkhaus Zagzebski, Linda. 2012. *Epistemic Authority: A Theory of Trust, Authority and Autonomy in Belief.* Oxford: Oxford University Press.

Tuck, Richard. 2009. *Free Riding.* Cambridge, MA: Harvard University Press.

Tuomela, Raimo. 2013. *Social Ontology: Collective Intentionality and Group Agents.* Oxford: Oxford University Press.

Turner, Chris. 2013. *The War on Science: Muzzled Scientists and Willful Blindness in Stephen Harper's Canada.* Vancouver: Greystone Books.

United Nations Environment Program. 2011. 'Thawing of Permafrost Expected to Cause Significant Additional Global Warming, Not Yet Accounted for in Climate Predictions.' <http://www.unep.org/newscentre/default.aspx?DocumentID=2698&ArticleID=9338>. Accessed 13 June 2013.

Vanderheiden, Steve. 2008. *Atmospheric Justice: A Political Theory of Climate Change.* Oxford: Oxford University Press.

Velleman, David. 2006. *Self to Self: Collected Essays.* Oxford: Oxford University Press.

Vogel, Stephen. 2012. 'Alienation and the Commons.' In *Ethical Adaptation to Climate Change: Human Virtues of the Future,* edited by Allen Thompson and Jeremy Bendik-Keymer, pp. 299–316. Cambridge, MA: The MIT Press.

Volk, Tyler. 2008. *CO_2 Rising: The World's Greatest Environmental Challenge.* Cambridge, MA: The MIT Press.

Waller, James. 2007. *Becoming Evil: How Ordinary People Commit Genocide and Mass Killing,* 2nd edition. Oxford: Oxford University Press.

Ward, Peter D. 2007. *Under a Green Sky: Global Warming, The Mass Extinctions of the Past, and What They Can Tell Us About our Future.* New York: Harper-Collins.

Weber, E. 2006. 'Experience-based and description-based descriptions of long-term risk: why global warming does not scare us (yet).' *Climatic Change* 77: pp. 103–20.

Weinberg, Rivka. 2008. 'Identifying and Dissolving the Non-Identity Problem.' *Philosophical Studies* 137 (1): pp. 3–18.

Weintrobe, Sally. 2013. 'The Difficult Problem of Anxiety in Thinking about Climate Change.' In *Engaging with Climate Change: Psychoanalytic and Interdisciplinary Perspectives*, edited by Sally Weintrobe, pp. 33–47. London: Routledge.

Weitzman, Martin L. 2009. 'On Modeling and Interpreting the Economics of Catastrophic Climate Change.' *Review of Economics and Statistics* 91 (1): pp. 1–19.

White, Lynn. 1974. *Ecology and Religion in History*. New York: Harper-Collins.

Wijkman, Anders and Johan Rocktröm. 2012. *Bankrupting Nature: Denying our Planetary Boundaries*. London: Routledge.

Wike, Richard. 31 March 2014. 'Many around the world see climate change as a major threat.' PEW Research Centre. <http://www.pewresearch.org/fact-tank/2014/03/31/many-around-the-world-see-climate-change-as-a-major-threat/>. Accessed 15 April 2014.

Williams, Bernard. 1985. *Ethics and the Limits of Philosophy*. London: Routledge.

Williams, Bernard. 1993. *Shame and Necessity*. Berkeley: University of California Press.

Williams, Bernard. 2002. *Truth and Truthfulness*. Princeton: Princeton University Press.

Williston, Byron. 2006. 'Blaming Agents in Moral Dilemmas.' *Ethical Theory and Moral Practice* 9: pp. 563–76.

Williston, Byron. 2012a. 'Climate Change and Radical Hope.' *Ethics and the Environment* 17 (2): pp. 165–85.

Williston, Byron. 2012b. 'The Importance of Self-Forgiveness.' *American Philosophical Quarterly* 49 (1): pp. 67–80.

Williston, Byron. 2011. 'Moral Progress and Canada's Climate Failure.' *Journal of Global Ethics* 7 (2): pp. 149–60.

Williston, Byron. 2002. 'Self-Deception and the Ethics of Belief: Locke's Critique of Enthusiasm.' *Philo* 5 (1): pp. 62–83.

Woodman. J. 1995. 'Considerations on the Keeping of Negroes.' *Portable Enlightenment Reader*, edited by Isaac Kramnick, pp. 630–7. London: Penguin.

Woodruff, Paul. 2011. *The Ajax Dilemma: Justice, Fairness and Rewards*. Oxford: Oxford University Press.

Woolard, Fiona. 2012. 'Have We Solved the Non-Identity problem?' *Ethical Theory and Moral Practice* 15 (5): pp. 677–90.

World Bank. 2013. *Turn Down the Heat: Climate Extremes, Regional Impacts and the Case for Resilience*. <http://www.worldbank.org/en/topic/climatechange/publication/turn-down-the-heat-climate-extremes-regional-impacts-resilience>. Accessed September 2014.

Wright, Ronald. 1997. *A Scientific Romance*. Toronto: Vintage Canada.

Yergin, Daniel. 2008. *The Prize: An Epic Quest for Oil, Money and Power*. New York: Simon and Schuster.

Yergin, Daniel. 2012. *The Quest: Energy, Security and the Remaking of the Modern World*. New York: Penguin.

Zografos, Christos et al. 2014. 'Sources of Human Insecurity in the Face of Hydro-Climatic Change.' *Global Environmental Change* (29), 327–36. <http://www.sciencedirect.com/science/article/pii/S0959378013001933>. Accessed 14 November 2014.

Index